Studies in Computational Intelligence

Volume 763

T0155563

Series editor

Janusz Kacprzyk, Polish Academy of Sciences, Warsaw, Poland
e-mail: kacprzyk@ibspan.waw.pl

About this Series

The series "Studies in Computational Intelligence" (SCI) publishes new developments and advances in the various areas of computational intelligence—quickly and with a high quality. The intent is to cover the theory, applications, and design methods of computational intelligence, as embedded in the fields of engineering, computer science, physics and life sciences, as well as the methodologies behind them. The series contains monographs, lecture notes and edited volumes in computational intelligence spanning the areas of neural networks, connectionist systems, genetic algorithms, evolutionary computation, artificial intelligence, cellular automata, self-organizing systems, soft computing, fuzzy systems, and hybrid intelligent systems. Of particular value to both the contributors and the readership are the short publication timeframe and the world-wide distribution, which enable both wide and rapid dissemination of research output.

More information about this series at http://www.springer.com/series/7092

Lídice Camps Echevarría • Orestes Llanes Santiago
Haroldo Fraga de Campos Velho
Antônio José da Silva Neto

Fault Diagnosis Inverse Problems: Solution with Metaheuristics

Springer

Lídice Camps Echevarría
Centro de Estudios de Matemática
Universidad Tecnológica de La Habana José
Antonio Echeverría, CUJAE
Marianao, La Habana, Cuba

Haroldo Fraga de Campos Velho
National Institute for Space Research, INPE
São José dos Campos
São Paulo, Brazil

Orestes Llanes Santiago
Dpto. de Automática y Computación
Universidad Tecnológica de La Habana José
Antonio Echeverría, CUJAE
Marianao, La Habana, Cuba

Antônio José da Silva Neto
Instituto Politécnico
Universidade do Estado do Rio de Janeiro,
UERJ
Nova Friburgo, Rio de Janeiro, Brazil

ISSN 1860-949X   ISSN 1860-9503   (electronic)
Studies in Computational Intelligence
ISBN 978-3-030-07908-6   ISBN 978-3-319-89978-7   (eBook)
https://doi.org/10.1007/978-3-319-89978-7

Printed on acid-free paper

This Springer imprint is published by the registered company Springer International Publishing AG part
of Springer Nature.
The registered company address is: Gewerbestrasse 11, 6330 Cham, Switzerland

To my mother Luz Milagros, in memoriam
Lídice Camps Echevarría

To Xiomara, Maria Alejandra, and Maria Gabriela
Orestes Llanes Santiago

To Nádia
Haroldo Fraga de Campos Velho

To Gilsineida, Lucas, and Luísa
Antônio José da Silva Neto

Preface

Faults can occur in every system, but when they occur in control systems, they may cause not only economic losses but also damage to human being, material, and to the environment. But, how is a fault in a control system defined? A fault is defined *as an unpermitted deviation of at least one characteristic property or parameter of a system from the acceptable, usual, or standard operating condition* [57, 61, 78, 115]. The area of knowledge related to methods for diagnosing faults is called Fault Diagnosis or Fault Detection and Isolation (FDI). The name Fault Detection and Isolation indicates how are conceived many Fault Diagnosis methods: first detecting if there are faults that are affecting the system and then isolating them, i.e., deciding which fault is affecting the system.

On diagnosing a system relies a big interest. Topics so important in the industry such as reliability, safety, and efficiency are related to FDI. Fault Diagnosis is also crucial for the topic of maintenance. It is also recognized that the increase of the automation in the industry enhances the probability of faults occurrence [61]. Therefore, it is easy to conclude that Fault Diagnosis is a current area of intense research and real applications interest. But FDI is not a simple and easy problem to formulate and solve. For diagnosing faults in systems, measurements need to be used. These measurements are commonly corrupted by noise. Moreover, the measurements can be affected by spurious disturbances acting on the system. All these facts can lead to the conclusion that the system may be deviated from its acceptable behavior, even when no faults are affecting it. This implies in false alarms. In order to avoid this situation, the methods developed and applied for FDI should be *robust*. Robustness means rejection of false alarms, which are attributable to disturbances or spurious signals. With the increase of the robustness it could occur, that some faults can not be diagnosis, for example when the fault effects in the system output are within the range of deviation due to the noise. This fact is related to the system loss of *sensitivity*. Therefore, useful FDI methods must present two properties: *robustness* and *sensitivity*. Moreover, in complex systems, the propagation of faults can rapidly occur [57, 115]. As a consequence, Fault Diagnosis also take into consideration the diagnosis time [57, 115].

A great variety of FDI methods can be found in the literature, and they may be brought down to two types of methods: model-based methods and non-model-based methods. Model-based methods make use of a mathematical or a physical/mathematical model of the system.

In this book, it is presented and formalized a recent methodology for Fault Diagnosis which falls into the category of model-based methods. The name of the methodology is *Fault Diagnosis—Inverse Problem Methodology* (FD-IPM). It unifies the results of some years of cooperation among researchers from the fields of FDI, Inverse Problems, and Optimization. Some of the contributions that resulted from this cooperation are described in Refs. [1, 18–21]. As its name indicates, it is based on the formulation of Fault Diagnosis as an Inverse Problem. In particular, the Fault Diagnosis Inverse Problems are formulated as optimization problems, which are solved using metaheuristics. Therefore, in the proposed methodology, the areas of FDI, Inverse Problems, and Metaheuristics are linked.

The main objective of this book is to formalize, generalize, and present in a systematic, organized, and clear way the main ideas, concepts, and results obtained during the last years which are based on the formulation of Fault Diagnosis as Inverse Problems which are solved as optimization problems.

For readers familiarized with Inverse Problems, some questions could arise: why formulate FDI as an Inverse Problem, if it is well known that Inverse Problems are usually ill-posed problems (sometimes the ill- posed problems are called hard problems), there are already well established FDI methods, which are not based on the Inverse Problem approach?

It is true that Inverse Problems are usually not easy to solve. The main difficult point to deal with, when solving them, is the effect on the solution when the observed data (e.g., measurements) present noise, i.e., amplification of the noise which is always present in the observable variable affects the Inverse Problem solution. Inverse problems belong to the class of ill-posed problems: the existence or uniqueness of the solution are not guarantee; or it does not show continuous dependence on the input values. From the latter condition, the noise in the measurements can be amplified in the inverse solution. But, in turn FDI also deals with noisy data (measurements of the system), and it is very important to know how the noise in the data influences the diagnosis of the faults, in order to avoid, for example, false alarms or to identify faults that may be masked within the noise. It is recognized that obtaining robust and sensitive FDI methods, at the same time, continues to be a current research field of high interest [114]. Moreover, Inverse Problems is an interdisciplinary area that matches the mathematical model of a problem with its experimental data [94]. That is exactly the idea behind FDI, when a mathematical or physical/mathematical model of the system is considered known.

Formulating FDI as an Inverse Problem also brings some insight to the understanding of the FDI problem by means of the introduction of ideas and methodologies from Inverse Problems. Inverse Problems, which arise from practical applications, deal with observable information from the system. How this information can be used is a key issue for solving Inverse Problems. Fault Diagnosis can be understood as a problem based on information (e.g., model of the system and measurements). Therefore, the ideas from Inverse Problems allowed to identify some results that could help solve problems of the area of Faults Diagnosis.

Moreover, the formulation of the Fault Diagnosis Inverse Problems as optimization problems allows to apply metaheuristics for computing the solution, i.e., diagnosing the faults.

Metaheuristics are a group of nonexact algorithms that allow to solve optimization problems based on a search strategy in the feasible solution space. Metaheuristics may provide a sufficiently good solution to an optimization problem, especially with incomplete information. Naturally an obvious question comes to mind. Which is the best metaheuristic for Fault Diagnosis Inverse Problems? In [137], it is demonstrated that there is not a unique answer to this question. Instead, this book describes in an easy way, but with rigor, four well-known metaheuristics for optimization (Differential Evolution, Particle Collision Algorithm, Ant Colony Optimization, and Particle Swarm Optimization), which were also used during the application of the methodology to Fault Diagnosis in three benchmark problems (DC Motor, Inverted Pendulum System, and Two Tanks System) from the FDI area. These examples are useful for showing how to analyze and interpret the influence of some metaheuristics parameters in the quality of the diagnosis. Furthermore, this book presents the main ideas and concepts from optimization and metaheuristics, which may be useful for readers that are not familiarized with these topics.

The formulation of FDI as Inverse Problems also allowed to develop two new hybrid metaheuristics: *Particle Swarm Optimization with Memory* and *Differential Evolution with Particle Collision*. New metaheuristics constantly arise in the literature, but for being accepted by the computational community, they have to be formalized and validated, following the well-recognized methodology for metaheuristic validations [30, 45, 125]. This is also presented in this book.

The analysis of the results obtained along the experiments for the validation of the new metaheuristics, as well as in their application to the benchmark problems, made use of basic concepts and test from statistics. During the analysis of the experiments, different tables and graphics were constructed in order to facilitate the presentation of the results, as well as their interpretation.

Formulating Fault Diagnosis as an Inverse Problem and solving it by means of metaheuristics bring together readers from at least three different areas: Fault Diagnosis, Inverse Problems, and Metaheuristics. This represents the nature of the authors of this book.

The prerequisites to read this book are calculus of several variables and linear algebra. Some basics about programming are also useful for a better understanding of the chapter that presents the topic of metaheuristics.

The chapters of this book are summarized as follows:

- **Chapter 1:** In this chapter, the main concepts, ideas, advantages, and disadvantages related to the use of model-based FDI methods are presented. The main ideas on the formulation and solution of Inverse Problems are also briefly described.

- **Chapter 2:** This chapter presents and formalizes Fault Diagnosis as an Inverse Problem, as well as the new methodology for Fault Diagnosis: Fault Diagnosis—Inverse Problem Methodology. The three benchmark problems used during the experiments are also described.
- **Chapter 3:** This chapter makes an introduction to metaheuristics for optimization. In particular, the metaheuristics Differential Evolution, Particle Collision Algorithm, Ant Colony Optimization for continuous problems, and Particle Swarm Optimization are described. This chapter also presents two new metaheuristics: Particle Swarm Optimization with Memory and Differential Evolution with Particle Collision.
- **Chapter 4:** This chapter presents the application of Fault Diagnosis—Inverse Problem Methodology to the three benchmark problems considered. In particular, the experiments are designed in order to analyze robustness and sensitivity of the diagnosis obtained with FD-IPM.

Chapter 1 can be read independently when the reader is only interested in the main concepts and ideas from model-based Fault Diagnosis or Inverse Problems. For readers interested in metaheuristics, we recommend Chap. 3 which can be read independently. For the new methodology, Chap. 2 has to be read. We recommend to read Chap. 1 before reading Chap. 2. Chapter 4 presents the applications of the methodology to three benchmark problems. Before reading Chap. 4, we recommend reading Chap. 2.

Appendices A and B show the Matlab$^{®}$ codes for the algorithms of the new metaheuristics Differential Evolution with Particle Collision and Particle Swarm Optimization with Memory, respectively.

La Habana, Cuba   Lídice Camps Echevarría
La Habana, Cuba   Orestes Llanes Santiago
São José dos Campos, Brazil   Haroldo Fraga de Campos Velho
Nova Friburgo, Brazil   Antônio José da Silva Neto
May 2018

Acknowledgments

The authors acknowledge the Springer representation in Brazil for the publication of this book. Our gratitude is also due to the Brazilian Research supporting agencies CAPES—Fundação Coordenação de Aperfeiçoamento de Pessoal de Nível Superior, CNPq—Conselho Nacional de Desenvolvimento Científico e Tecnológico, and FAPERJ—Fundação Carlos Chagas Filho de Amparo à Pesquisa do Estado do Rio de Janeiro, as well as to the Cuban Ministry of Higher Education MES (Ministerio de Educación Superior).

Orestes Llanes Santiago acknowledges specially CNPq-Conselho Nacional de Desenvolvimento Científico e Tecnológico for the Special Visiting Professor grant No.401023/2014-1.

Contents

Acronyms

ABC	Artificial Bee Colony
ACO	Ant Colony Optimization
BBO	Biogeography-Based Optimization
BMs	Bio-Inspired Metaheuristics
CS	Cuckoo Search
DC	Direct Current
DE	Differential Evolution
DEwPC	Differential Evolution with Particle Collision
EAs	Evolutionary Algorithms
EDAs	Estimation of Distribution Algorithms
EP	Evolutionary Programming
FA	Firefly Algorithm
FDI	Fault Detection and Isolation, Fault Diagnosis
FD-IPM	Fault Diagnosis—Inverse Problem Methodology
GAs	Genetic Algorithms
GEO	Generalized Extremal Optimization
GP	Genetic Programming
IP	Inverse Problems
IPS	Inverted Pendulum System
IWD	Intelligent Water Drops Algorithm
LTI	Linear Time Invariant
MAs	Memetic Algorithms
MPCA	Multiple Particle Collision Algorithm
NMs	Nature-Inspired Metaheuristics
NPMs	Non-population-Based Metaheuristics
PCA	Particle Collision Algorithm
PI	Proportional-Integral Controller
PID	Proportional-Integral-Derivative Controller
PMs	Population-Based Metaheuristics
PSO	Particle Swarm Optimization
PSO-M	Particle Swarm Optimization with Memory

SA	Simulated Annealing
SI	Swarm Intelligence
SIs	Swarm Intelligence algorithms
SISO	Single Input Single Output
TS	Tabu Search

Chapter 1
Model Based Fault Diagnosis and Inverse Problems

This chapter introduces the main concepts and ideas concerning Model based Fault Diagnosis in Sect. 1.1. Fault Diagnosis based on parameter estimation is presented in Sect. 1.2 and Inverse Problems are briefly introduced in Sect. 1.3.

1.1 Model Based Fault Diagnosis Methods

The first Fault Diagnosis methods were introduced at the beginning of the 60s of the twentieth century. They were based on monitoring certain observable parameters. The objective was to verify if the values of such parameters were within certain acceptable ranges [57]. Those methods do not allow the early detection of faults, which slowly affect the monitored parameters. Therefore, the system cannot be diagnosed [57, 60].

Many of the initial works in Fault Detection and Isolation (FDI) were linked with chemical processes. In 1978, D. M. Himmeiblau published the first book about model based FDI methods and their applications to chemical processes [64].

Important works that intended to unify the first results in the FDI are [42, 55, 59, 64]. In 1991 was created the SAFEPROCESS (Fault detection, supervision and safety for technical processes) Committee as part of the *International Federation of Automatic Control*, IFAC.

Nowadays, the FDI methods are divided into two main groups: those which are based on a mathematical or physical model of the system to be diagnosed, and those which do not use a model [130–132]. The methods of the first group are called Analytical Methods or simply Model based Methods. The methods of the second group are known as historical data or data driven based methods.

Within the historical data based methods, it is possible to find methods based on statistics, pattern recognition, heuristic knowledge, and artificial intelligence [16, 53,

© Springer International Publishing AG, part of Springer Nature 2019
L. Camps Echevarría et al., *Fault Diagnosis Inverse Problems: Solution with Metaheuristics*, Studies in Computational Intelligence 763,
https://doi.org/10.1007/978-3-319-89978-7_1

109, 120]. All these methods have, as main disadvantage, the need of big amount of historical measurements of the system, which should contain information about all the faults to be diagnosed [2].

The structure of model based FDI methods can be represented by the general scheme shown in Fig. 1.1. The idea behind these methods is the generation of residuals to be used for diagnosing the system. These residuals are generated based on the input, output, and a model of the system [42]. For this purpose, the system is usually divided into three main parts to be diagnosed: actuators, process, and sensors. As a consequence, the faults are also classified considering the part of the system that they affect, as [31, 42, 61]:

- f_u: faults in actuators
- f_p: faults in process
- f_y: faults in sensors

In Fig. 1.2 it is represented the classification of the faults used throughout this book.

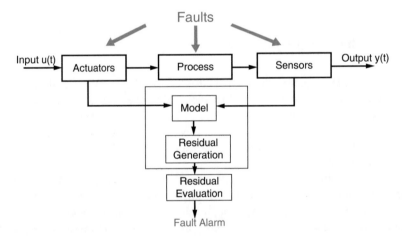

Fig. 1.1 General structure of model based FDI methods

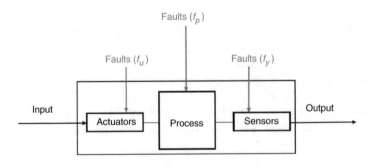

Fig. 1.2 Classification of faults considering the part of the system that they affect

The faults are also classified based on their nature. There are incipient faults, abrupt faults, constant faults, intermittent faults, and faults that combined more than one kind of fault.

The model based methods are divided into three main groups, which correspond to three main approaches: observers; parity space; and parameters estimation [55, 57, 58, 64, 76, 96].

The methods based on observers were introduced for linear systems during the 70s of the twentieth century. In [134] it is presented a summary about these first steps. Parity Space was introduced during the 80s and they were formally established in [99]. Some works that show the relationship among these methods are in Refs. [24, 31, 41, 61, 115, 130, 136]. FDI based on parameters estimation was formalized in 1987, by H. Hohmann, in his doctoral thesis [56]. Isermann [57] summarizes about FDI based on parameters estimation.

Despite the fact that these three approaches have been extensively studied and developed, their limitations are also well recognized. For example, in [41, 136], it is presented a clear description of each approach and their limitations:

- Methods based on observers: They do not always allow the isolation of actuators faults. In the case of nonlinear systems, the complexity on designing the observer increases. As a consequence, its applications are focused on laboratory systems.
- Methods based on Parity Space: They also do not allow the isolation of actuators faults in many situations. In the case of nonlinear systems, it is needed an exact model of the system.
- Methods based on Parameters Estimation: They demand the knowledge of the relation *faults/physical coefficients of the system/model parameters*. They are not useful for diagnosing sensors faults. Their computational costs are high. In case of not having a one-to-one relationship between model parameters and system parameters, the isolation of faults is very difficult or not possible. It is recognized that these methods have also limitations for detecting actuators faults.

Recent techniques for model based FDI are mostly a variation of methods based on observer or parity space, with the introduction of other techniques (e.g., classification methods from Bayes classification, neural networks with decision trees, inference methods from approximate reasoning with fuzzy logic and hybrid neuro-fuzzy) [62]. Therefore, they maintain the main limitations of the original methods [40, 80, 95, 96]. In [63] it is presented the application of the basic methods of FDI to selection of twenty real technical components and processes as examples.

In FDI, there are two main desirable characteristics to be reached:

- **Robustness:** The systems can be affected by internal or external disturbances. Moreover, they are usually immersed in noisy environments. As a consequence, the impact of these situations can be confused with the effect of faults. Therefore, desirable FDI methods should be able to handle these situations in a proper way. Meaning they should be *robust*.
- **Sensitivity:** With the aim of increasing the robustness, FDI methods could not able to detect small magnitude faults. For example, those whose effects in the

system output are within the range of deviation due to the noise in the system. Therefore, another desirable characteristic of the FDI methods is the sensitivity for diagnosing small magnitude faults.

A proper balance between robustness and sensitivity, as well as a proper diagnosis time (latency time) which allows the on-line diagnosis, is the key of success measuring to FDI [31, 114, 115].

In order to obtain robust and sensitive diagnosis, the FDI methods based on observers or parity space make additional efforts for generating robust and sensitive residuals. Due to the fact that the residual generation depends on the kind of model used to describe the system, and on the kind of faults to be diagnosed, these schema for residual generation do not allow an easy generalization [41, 136].

Therefore, obtaining robust and sensitive FDI methods continues to be a current research field of high interest [114, 115].

Within the computational intelligence framework, the fuzzy logic and neural networks have been widely applied in model based and historical data-based FDI methods [29, 44, 88, 135, 142]. On the other hand, just a few papers have reported applications of metaheuristics to FDI [41, 66, 82, 106, 133, 135, 143].

Witczak [136] presents a state of the art of FDI, considering also the application from the computational intelligence area. Applications of evolutionary algorithms and neural networks to FDI are worked in [135].

1.2   Fault Diagnosis Methods Based on Parameter Estimation

FDI based on model parameters, which are partially unknown, requires online parameters estimation methods [57–59, 61]. For that purpose, the input vector $\mathbf{u}(t) \in \mathbb{R}^m$, the output vector $\mathbf{y}(t) \in \mathbb{R}^p$, and the basic model structure must be known [61].

The models for describing the systems depend on the dynamics of the process, and the objective to be reached with the simulation. These models can be linear or nonlinear, and can be represented or not in the state space. The most used model is the linear time invariant (LTI), which has two representations: the transfer function or transfer matrix, and the state space representation [97]. This latter representation is also valid for nonlinear models [97].

Let's consider a state space representation of a controlled system [61]:

$$\begin{aligned}
\dot{\mathbf{x}}(t) &= f\left(\mathbf{x}(t), \mathbf{u}(t), \mathbf{\Theta}\right) \\
\mathbf{y}(t) &= g\left(\mathbf{x}(t), \mathbf{u}(t), \mathbf{\Theta}\right)
\end{aligned} \tag{1.1}$$

where the vector of state variables is represented by $\mathbf{x}(t) \in \mathbb{R}^n$. The measurable input signals $\mathbf{u}(t) \in \mathbb{R}^m$ and output signals $\mathbf{y}(t) \in \mathbb{R}^p$ may be directly obtained by the use of physical sensors; and $\mathbf{\Theta} \in \mathbb{R}^j$ is the model parameters vector.

The LTI models can be represented by the following equations:

$$\dot{x}(t) = Ax(t) + Bu(t)$$
$$y(t) = Cx(t) + Du(t)$$

(1.2)

being A, B, C, and D matrices with appropriate dimensions: $A \in \mathbb{M}_{n \times n}(\mathbb{R})$, $B \in \mathbb{M}_{n \times m}(\mathbb{R})$, $C \in \mathbb{M}_{p \times n}(\mathbb{R})$, and $D \in \mathbb{M}_{p \times m}(\mathbb{R})$.

For the case of a model based on the input/output behavior of SISO (single input single output) processes by means of ordinary linear differential equations, its vectorial form is represented as:

$$y(t) = \psi(t)\Theta^T(t)$$

(1.3)

being

$$\Theta(t) = (a_1 \dots a_n \ b_0 \dots b_m)$$

(1.4)

i.e., $j = n + m + 1$, and

$$\psi(t) = \left(-y^{(1)}(t) \dots - y^{(n)}(t) \ -u^{(1)}(t) \dots - u^{(m)}(t) \right)$$

(1.5)

where $y^{(n)}(t)$ and $u^{(m)}(t)$ indicate derivatives:

$$y^{(n)}(t) = \frac{d^n y(t)}{dt^n}$$

(1.6)

and

$$u^{(m)}(t) = \frac{d^m u(t)}{dt^m}$$

(1.7)

The respective transfer function becomes, through Laplace transformation:

$$G_p = \frac{y(s)}{u(s)} = \frac{B(s)}{A(s)} = \frac{b_m s^m + b_{m-1}s^{m-1} + \dots + b_1 s + b_0}{a_n s^n + a_{n-1}s^{n-1} + \dots + a_1 s + 1}$$

(1.8)

The components of the model parameter vector can be identified with components of the physical process coefficients vector $\rho \in \mathbb{R}^r$. The variations of these coefficients can also be associated with faults. In this situation, the diagnosis of faults can be obtained using the estimations of these process parameters once the relationships between $\Theta - \rho$ and $\rho -$ **faults** are established [57].

This divides the diagnosis based on parameter estimation into two steps. The first one considers the estimation of the parameter vector Θ, permitting the detection; and the second includes isolation and identification of faults based on the mentioned relationships [57].

If $j \leq r$, then the relationship between Θ and ρ will be not one to one. In that case, some faults will not be isolated from the rest, which means that some faults cannot be eventually isolable from at least one other fault [42, 43, 57, 61]. This is one of the drawbacks of this method; the process of fault isolation may become extremely difficult because model's parameters do not uniquely correspond to those of the system under analysis.

The two methods that are commonly used for estimating the model parameters Θ are classified with respect to the minimization function that they use [42, 61]:

1. **Least squares of the summation of the equation errors:** These methods require numerical optimization methods. They give more precise parameter estimations, but the computational effort is higher. Therefore, they are in general not suitable for on-line applications [61].
2. **Least squares of the summation of the output errors:** They usually need the use of filters in order to improve the estimations [61].

Therefore, it can be concluded that one feature of FDI based on parameter estimation is that the model parameters should have physical meaning. This approach is only recommended for the case of faults having a direct influence on the model's parameters, causing lack of success for cases of sensor and actuators faults [89, 136].

A drawback of FDI based on parameter estimation is the resulting high computational time [42, 58, 61, 135, 136]. This brings difficulties in applications for real on-line processes [61]. It is also added a typical limitation regarding parameter estimation-based approaches: the input signal should be persistently excited [89, 136].

1.3   Inverse Problems

A direct problem considers known: the phenomena involved in the system under study; their physical or mathematical model; and a certain *cause* (e.g., initial conditions, boundary conditions, and internal sources/sinks). It is then possible to simulate or describe the effect that the given cause provokes [111–113]. It means that the problem can be solved.

On the other hand, in Inverse Problems, the model of the system and its output are known or partially known, and it is necessary to determine the causes that yield such output (the direct problem effects) [17, 84, 85, 94, 111–113, 117]. A typical example is when for a given set of numerical values or measurements, which correspond to the solution of a direct problem, it is necessary to determine unknown initial conditions, boundary conditions, model parameters, or source/sink functions.

A widely cited general description of Inverse Problems is: *We call two problems inverses of one another if the formulation of each involves all or part of the solution of the other. Often, for historical reasons, one of the two problems has been studied extensively for some time, while the other is newer and not so well understood. In*

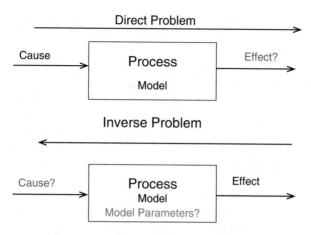

Fig. 1.3 Relation between direct and inverse problems [111]

such cases, the former problem is called the direct problem, while the latter is called the Inverse Problem [68]. This description shows that both problems, the direct and the inverse, are connected.

There are several definitions for Inverse Problem [17]. One definition can be stressed as:

Definition 1.1 Solving an ***Inverse Problem*** is to determine unknown causes from observed or desired effects *[17, 39]*

Figure 1.3 shows a scheme, which represents the relation between a direct (forward) problem and its Inverse Problem.

Considering the causes to be determined, Inverse Problems are classified into three main groups: parameter estimation inverse problems, inverse problems for estimating initial conditions, and inverse problems for estimating boundary conditions [39]. In [9] the inverse problems are classified considering the dimension (finite or not) of the *cause* to be determined: parameter estimation inverse problems (finite dimension) and function estimation inverse problems (infinite dimension).

When studying Inverse Problems, there is an important definition which was introduced in 1923 by Hadamard, the well-posed problems by Hadamard:

Definition 1.2 Let's $Ax = f$ with $x \in X$; $f \in F$; and A a differential operator with the domain of definition D_A. The problem is ***well-posed***, in the Hadamard's sense, if the three following conditions are satisfied [51]:

1. Existence of the solution: $\forall f \in F \longrightarrow \exists x \in D_A : Ax = f$
2. Uniqueness of the solution: $\forall x_1, x_2 \in D_A : Ax_1 = Ax_2 = f \longrightarrow x_1 = x_2$
3. Stability: $\forall x_1, x_2 \in D_A : Ax_1 = f_1, Ax_2 = f_2$ if $x_1 \to x_2$ in $X \longrightarrow f_1 \to f_2$ in F

If a problem does not satisfy any of the three above conditions, then it is an ill-posed problem.

The first two conditions in Definition 1.2 are related with the mathematical determination in the sense that they demand that the solution exists, and if it is determined by non-numerical methods, it can be uniquely determined. The third condition is related with the numerical determination, which means that if the solution is computed by numerical methods, the stability condition is needed.

In many Inverse Problems, small variations in the data do not necessarily imply small variations in the causes to be determined. Therefore, a difficult point, when solving an Inverse Problem, is the effect on the solution when the observed data (e.g., measurements) present noise, i.e. amplification of the noise present in the observable variable affects the inverse problem solution. Inverse Problems are ill-posed.

If the observable data in the Inverse Problem is defined as a solution or a set of solutions to the direct problem, then the first condition in Definition 1.2 is clearly satisfied. But in many Inverse Problems, the existence of the solution can fail if the observable data is corrupted with noise, which is a typical situation in Inverse Problems that arise from real applications [6].

If condition 2 from Definition 1.2 is not satisfied, the problem should be reformulated by means of incorporating additional observable data; or restricting the set of feasible solutions. For restricting the set of feasible solutions, *a priori* information about the problem is usually needed [6].

For Inverse Problems, which the existence and uniqueness of their solutions is not clear and there are enough observable noisy data, it is useful to search an approximate solution of the Inverse Problem instead the exact one, by means of the minimization of a certain measure of discrepancy between the data without noise that can be obtained from the direct model and the observable data corrupted by noise [94].

If the third condition from Definition 1.2 is not satisfied, it can occur that noise present in the measurements and round-off errors during computations can be amplified by an arbitrarily large factor, and make a computed solution completely useless to deal with [6]. This is a typical situation when dealing with Inverse Problems and make them specially interesting and difficult [94].

It can be concluded that an important point in Inverse Problems is determining the quantity and the quality of information about it. Another point is that in practice, it is usual to deal with nonexact data. Therefore, to have too much observable data is the best, even if some repetition occurs [94].

Concerning the methods for solving Inverse Problems, there are two main classifications [94, 112, 113]:

- Explicit methods: Methods, which are obtained through direct inversion of the operator that represents the direct model problem.
- Implicit methods: Iterative methods, which explore the space of feasible solutions of the Inverse Problem until a certain criterion is reached. It means, that they usually look for a good enough approximated solution of a certain Inverse Problem.

The implicit methods for Inverse Problems are related to the formulation of Inverse Problems as optimization problems, usually involving the minimization of an objective function. The optimization problems arise when a certain measure of discrepancy between the data without noise, which can be obtained from the direct model and the observable data, corrupted by noise, is minimized in order to find an approximated solution to the Inverse Problem. For these methods, it is usual to have as the objective function, the sum of the quadratic errors between the observations and the calculated values for such observations by means of the direct problem model [5]. Therefore, the solution of the Inverse Problem involves the solution of the direct problem model. The formulation as an optimization problem addresses the existence and uniqueness of the Inverse Problem. For solving the optimization problem, it is usual to use computational methods, in particular metaheuristics for optimization [74, 84, 85, 104, 105, 111, 112]. These metaheuristics are usually applied to solving parameter estimation inverse problems and they are complemented with a sensitivity analysis of the output of the system with respect to the parameters to be estimated. As a result of the sensitivity analysis, it can be decided which parameters can be correctly estimated. Therefore, the sensitivity analysis is an alternative way for studying how the noise in the data can influence in the quality of the solution to the Inverse Problem under study [112].

As previously mentioned, a usual way to deal with an ill-posed inverse problem is to add more previous information about it, e.g. information concerning the derivatives of the solution of the Inverse Problem. When enough good *a priori* information about the solution of the Inverse Problem is available, it can be added to the objective function in the optimization problem. The most usual way to add this information is by means of a regularization term, which allows the Inverse Problem to become a well-posed problem. The regularization methods are therefore very useful [92, 94, 103, 116], being the Tikhonov Regularization the most common approach. Moreover, Tikhonov re-defined the conditions for well-posed problem. As a consequence, there are new conditions for defining a well-posed problem, in the Tikhonov's sense.

In recent years, it has been observed an increase in the use of statistical techniques for the formulation and solution of Inverse Problems [98, 113, 126–128], with special interest in the Bayesian Theory [93].

1.4 Remarks

In this chapter, the main concepts, ideas, advantages, and disadvantages related to the use of model based FDI methods are presented. The main ideas on the formulation and solution of Inverse Problems are also briefly described.

Chapter 2
Fault Diagnosis Inverse Problems

This chapter deals with FDI as an Inverse Problem. Section 2.1 presents and formalizes Fault Diagnosis as an Inverse Problem. In Sect. 2.1, it is also formalized a new methodology for Fault Diagnosis: Fault Diagnosis—Inverse Problem Methodology (FD-IPM). Section 2.2 describes the mathematical models to be used by FD-IPM. Section 2.3 describes an alternative approach for determining uniqueness in Fault Diagnosis Inverse Problems, which is based on the structural analysis of the model that describes the system.

Section 2.4 describes the three benchmark problems considered: DC Motor, Inverted Pendulum System, and Two Tanks System [31, 86] in Sect. 2.4.3. Section 2.5 shows the application of the first step of FD-IPM to each benchmark problem.

2.1 Formulation of Fault Diagnosis as an Inverse Problem and the Fault Diagnosis: Inverse Problem Methodology (FD-IPM)

The Fault Diagnosis Problem can be seen as an Inverse Problem: taking as known partial information on output and input of the system, as well as a mathematical or physical model that describes the system, one wants to determine the causes that affect the response of the system. In this case, the faults are the causes in the definition of Inverse Problem, see Definition 1.1 in Sect. 1.3. However, as described in Sect. 1.1, other causes may be affecting the system, e.g. disturbances or noise. In Fig. 2.1, it is represented the Fault Diagnosis Inverse Problem, and its direct problem.

© Springer International Publishing AG, part of Springer Nature 2019
L. Camps Echevarría et al., *Fault Diagnosis Inverse Problems: Solution with Metaheuristics*, Studies in Computational Intelligence 763, https://doi.org/10.1007/978-3-319-89978-7_2

Fig. 2.1 (**b**) Fault Diagnosis
Inverse Problem and its (**a**)
direct problem

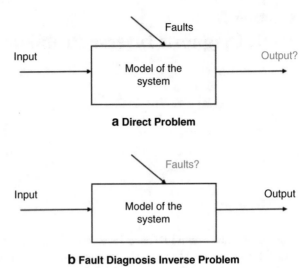

a Direct Problem

b Fault Diagnosis Inverse Problem

Despite the fact that FDI can be described as an Inverse Problem, this approach has not been intensively used. Some recent incursions have been reported in this sense [22, 143], but they were limited to a particular part of a system, and no generalizations are provided.

From the representation of the Fault Diagnosis Inverse Problem in Fig. 2.1, it is clear that a model that represents the system is needed. Moreover, this model of the system should incorporate the faults in an explicit way. For this purpose, it is used the fault classification that was shown in Sect. 1.1 (see Fig. 1.2). This fault classification is also used in other model based FDI methods [31, 42, 61]. Following this classification, a fault vector, represented by $\theta_f = (\mathbf{f_u}\ \mathbf{f_p}\ \mathbf{f_y})^T$ can be explicitly introduced in the model that describes the system. As a consequence, the estimation of the fault vector θ_f allows to diagnose the system. Therefore, the Fault Diagnosis Inverse Problem may be described as an Inverse Problem of parameter estimation, being θ_f the parameter to be estimated. Moreover, it can be formulated as an optimization problem: find the estimated value of θ_f that minimizes a certain *residual*. The estimated value for θ_f is denoted by $\hat{\theta}_f$.

Next are listed some of the key points addressed so far:

- The disadvantages of the current FDI model based methods are well recognized, see Sect. 1.1. It is also recognized that new FDI model based methods should be developed [114, 115].
- It is well known that the computational methods are suitable for solving Inverse Problems, specifically, the metaheuristics for optimization [111]. Some metaheuristics were reported to be successfully applied in the solution of FDI [35, 41, 82, 106, 133, 135, 143].
- Fault Diagnosis may be formulated as an Inverse Problem.

In this section, it is formalized a new methodology for FDI. It is called Fault Diagnosis—Inverse Problem Methodology (FD-IPM). This methodology is based on the direct estimation of a fault vector $\hat{\boldsymbol{\theta}}_f$, instead of estimating a model parameter vector $\boldsymbol{\Theta}$. For this, Fault Diagnosis is formulated as an Inverse Problem of parameter estimation. In order to solve the Fault Diagnosis Inverse Problem, it is formulated as an optimization problem, which is solved with the use of metaheuristics.

For solving the Inverse Problem of estimating $\boldsymbol{\theta}_f$, an optimization problem will be formulated. It is clear that the estimation of $\boldsymbol{\theta}_f$, whose result is denoted as $\hat{\boldsymbol{\theta}}_f$, should provide the minimal difference between the observations of the system (measurement of the output of the system under the effect of the fault vector $\boldsymbol{\theta}_f$) and the estimations of these observations when considering that the system is under the effect of a fault vector $\hat{\boldsymbol{\theta}}_f$. Therefore, the solution of the Fault Diagnosis Inverse Problem can be obtained by solving a minimization problem. Let' s note that obtaining the estimations for the output corresponds to solving the direct problem represented in Fig. 2.1, i.e. solving the model that describes the system with the estimated values for the fault vector $\hat{\boldsymbol{\theta}}_f$. The measurements of the output are usually obtained at certain instants of time, which are called sampling times.

Let's call I the number of sampling times; $Y_t \in \mathbb{R}$ contains the measurements of the output of the system to be diagnosed at a certain time t, with $t = 1, 2, \ldots, I$; $\hat{Y}_t \in \mathbb{R}$ are the estimated (calculated) values for the output of the system at each instant of time t, which are obtained from the solution of the direct problem for the system to be diagnosed. The output of the system should represent the effect of the faults that affect the system. Therefore, it can be written: $Y_t(\boldsymbol{\theta}_f)$ and $\hat{Y}_t(\hat{\boldsymbol{\theta}}_f)$. The estimations of the output are obtained solving the direct model of the system, taking as known as estimation value $\hat{\boldsymbol{\theta}}_f$ of the fault vector affecting the system. The optimization problem for solving the Fault Diagnosis Inverse Problem is formulated as:

$$\min \ f(\hat{\boldsymbol{\theta}}_f) = \sum_{t=1}^{I} \left[Y_t(\boldsymbol{\theta}_f) - \hat{Y}_t(\hat{\boldsymbol{\theta}}_f) \right]^2 \tag{2.1}$$

The solution of the optimization problem in Eq. (2.1) is an estimation of the fault vector affecting the system: $\hat{\boldsymbol{\theta}}_f$. The reliability of the diagnosis based on this formulation is intrinsically related with the quality of the estimation of the fault vector. The optimization problem in Eq. (2.1) corresponds to a system with only one output $Y_t(\boldsymbol{\theta}_f) \in \mathbb{R}$. Its generalization to systems with more than one output, i.e. $\boldsymbol{Y}_t(\boldsymbol{\theta}_f) \in \mathbb{R}^p$, leads to the following optimization problem:

$$\min \ f(\hat{\boldsymbol{\theta}}_f) = \left\| \sum_{t=1}^{I} \left[\boldsymbol{Y}_t(\boldsymbol{\theta}_f) - \hat{\boldsymbol{Y}}_t(\hat{\boldsymbol{\theta}}_f) \right]^2 \right\|_{\infty} \tag{2.2}$$

as in the optimization problem in Eq. (2.1), I represents the number of sampling times; $\boldsymbol{Y}_t(\boldsymbol{\theta}_f) \in \mathbb{R}^p$ contains the measurements of the output at a certain time t;

$\hat{Y}_t(\hat{\boldsymbol{\theta}}_f) \in \mathbb{R}^p$ are estimated (calculated) values for the output of the system at each instant of time t. For obtaining an objective function $f(\hat{\boldsymbol{\theta}}_f)$ with image in $\mathbb{R}$, it is introduced the infinite norm $\|g\|_\infty = \max\ g_i$.

For the solution of the minimization problem in Eq. (2.2), it is proposed the use of metaheuristics for optimization. For a better performance of the metaheuristics, it is convenient to add restrictions to the possible values of the fault vector to be estimated. Therefore, the optimization problem for the solution of the Fault Diagnosis Inverse Problem is reformulated as:

$$\min\ f(\hat{\boldsymbol{\theta}}_f) = \left\| \sum_{t=1}^{I} \left[Y_t(\boldsymbol{\theta}_f) - \hat{Y}_t(\hat{\boldsymbol{\theta}}_f) \right]^2 \right\|_\infty$$

$$\text{s.t.}\ \ \boldsymbol{\theta}_{f(\min)} \le \hat{\boldsymbol{\theta}}_f \le \boldsymbol{\theta}_{f(\max)} \tag{2.3}$$

The values $\boldsymbol{\theta}_{f(\min)}$ and $\boldsymbol{\theta}_{f(\max)}$ can be obtained with the help of experts, or analyzing the nature of the system to be diagnosed.

Let's note that when applying the metaheuristics for solving the optimization problem in Eq. (2.3), each evaluation of the objective function implies the solution of the direct problem in order to obtain the estimations $\hat{Y}_t(\hat{\boldsymbol{\theta}}_f) \in \mathbb{R}^p$. This can imply to solve a system of differential equations, for example. In Sect. 2.2, are described in detail, the models that can be used for representing the systems to be diagnosed.

In Fig. 2.2 is presented a schematic representation of the Fault Diagnosis Inverse Problem, formulated as an optimization problem.

For the application of the Fault Diagnosis—Inverse Problem Methodology (FD-IPM), **the system must satisfy the following three hypotheses:**

H1.   There is a known model of the system, $\mathfrak{M}$, which incorporates the dynamics of the faults in an explicit way, by means of the fault vector $\boldsymbol{\theta}_f \in \mathbb{R}^{(m+p+q)}$. This model represents the dynamics of a maximum of $(m + p + q)$ faults: where each

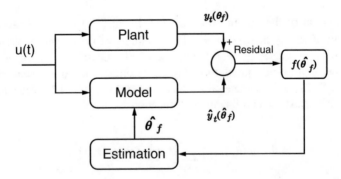

Fig. 2.2 Schematic representation of the Fault Diagnosis Inverse Problem, formulated as an optimization problem

of the m actuators and the p output sensors can be affected only by one fault at each time instant; and a total of q faults may occur in the in process.

H2. The faults are represented in the model as functions that do not change with time (constant functions); or functions that can be approximated by a constant function during the sampling time considered in the diagnosis.

H3. The system satisfies structural detectability and structural isolability (separability) for all faults.

Hypothesis H1 is related with the fact that FD-IPM is a model based methodology. Therefore, it needs a model of the system to be diagnosed. The kind of models that are needed for diagnosing with FD-IPM are described in Sect. 2.2. These models are also used in other model based methods [31, 42, 61]. They may be obtained by identification techniques or by recommendation of experts.

In hypothesis H2, it is important to know for which faults it is possible to consider the approximation by constant functions, without losing too much information. Lemma 2.1 next shows for which faults such approximation is adequate.

Lemma 2.1 *The approximation of a fault $f_1 = g_1(t)$ by a constant function will be better for faults that do not change in an abrupt way.*

Proof

Consider a fault $f_1 = g_1(t)$ which is not constant in time.

Consider t_0 the initial sampling time and $t_f = t_I = t_0 + I \Delta t$ the final time of sampling, being I the number of sampling instants; Δt the sampling interval; and $t_i = t_0 + i \Delta t$, with $i = 1, 2, \ldots, I$, are the time instants.

Consider $g_1(t)$ to be continuous in $[t_0, t_f]$, and also that

$$g_1(t) = \sum_{n=0}^{\infty} \frac{g_1^{(n)}(t_0)}{n!} (t - t_0)^n \text{ in } [t_0, t_f] \tag{2.4}$$

The function $f_1 = g_1(t)$ can be approximated in $[t_0, t_f]$ by Taylor Polynomials, in particular by the grade zero polynomial $f_1 = T_0 = g_1(t_0)$.

If $\left| \frac{dg_1}{dt}(t_i) \right| \leq M_{f1} \ \forall i = 1, 2, \ldots, I$, then the error of this approximation is:

$$R_0(t) \leq M_{f1}(t - t_0) \tag{2.5}$$

From expression (2.5), the error $R_0(t)$ is proportional to the value of M_{f1}. Therefore, for a fault f_1 such that $\frac{dg_1}{dt}(t_i) \approx 0$ (it means that its variation on time is almost zero), it is possible to approximate it by a constant function $f_1 = T_0 = g_1(t_0)$ with error $R_0 \approx 0$.

□

Hypothesis H2 is easy to verify, once a description of the system to be diagnosed is known. Let's note that in hypothesis H2 it is declared for which kind of fault, based on its nature, that the FD-IPM may be used.

Hypothesis H3 guarantees that all the faults affecting the system can be estimated based on the output of the system. It means that the Fault Diagnosis Inverse Problem of the system under study has a unique solution, see Sect. 2.3 for more details concerning this hypothesis and its verification.

The **Fault Diagnosis—Inverse Problem Methodology (FD-IPM)** is described as follows:

Step 1.   Verify hypothesis H1–H3. If the system satisfies the three hypotheses, formulate the optimization problem that corresponds to the Fault Diagnosis Inverse Problem for the system to be diagnosed; see optimization problem in Eq. (2.3), and go to **Step 3**. If the system satisfies H1 and H2, but does not satisfy H3 because the isolability requirement fails, then formulate the optimization problem that corresponds to the Fault Diagnosis Inverse Problem for the system to be diagnosed; see optimization problem in Eq. (2.3), and go to **Step 2**. Otherwise, the FD-IPM cannot be applied.

Step 2.   Detection: Take the vector $\hat{\boldsymbol{\theta}}_f = \mathbf{0}$ as a solution of the optimization problem obtained in **Step 1** and compute the objective function $f(\hat{\boldsymbol{\theta}}_f)$. If $f(\hat{\boldsymbol{\theta}}_f) < f_{umbral}$, where f_{umbral} is a relatively low value defined *a priori*, then the system is not under the influence of faults. Otherwise, the system is affected by faults.

Step 3.   Solving the Fault Diagnosis Inverse Problem: Apply metaheuristics for solving the optimization problem formulated in **Step 1**. Start the execution of the metaheuristics taking as the initial solution $\hat{\boldsymbol{\theta}}_f^{(0)} = \mathbf{0}$

Step 4.   Diagnosis: If any component of $\hat{\boldsymbol{\theta}}_f$, obtained in **Step 3**, is different from zero, then the fault that corresponds with to that specific component is affecting the system. The magnitude of the fault coincides with the value of its estimate.

Step 1 verifies the hypothesis for the application of the methodology, as well as formulates the optimization problem to be solved in order to diagnose the system. Step 2 is related to the detection of faults for those systems where the isolability of faults is not possible; Steps 3 and 4 are related to the detection and isolation of the faults for the systems that satisfy the three hypotheses H1, H2, and H3 of the FD-IPM. With the detection and isolation of faults, the diagnosis is complete.

When the system satisfies hypothesis H1 and H2, but does not satisfy hypothesis H3, the condition of detectability or isolability of the faults is not satisfied. Therefore, the system cannot be diagnosed. **Two situations can occur:**

- **Situation 1:** The system satisfies detectability and not the separability. In this case, the detection of the faults can be made, i.e. it is possible to determine if the system is under the effect of faults, but reliable estimations of the fault vector cannot be obtained. This detection is made with the help of Step 2 of the FD-IPM.
- **Situation 2:** The system does not satisfy detectability (as a consequence the separability is also not satisfied). In this case, the detection of the faults is not possible.

For systems in Situation 1, Step 2 is important. In such systems, the diagnosis cannot be made, but it can be determined if the system is affected or not by faults, even when the fault values cannot be estimated, and it cannot be decided which fault is affecting the system. This is made in Step 2, by computing the objective function of the optimization problem formulated in Step 1, considering that no faults are affecting the system, i.e. taking initially $\hat{\boldsymbol{\theta}}_f = \mathbf{0}$. If the value of the objective function is zero, it means that there are no differences between the measurements of the system output and the computed output when solving the direct problem of the system under study considering $\hat{\boldsymbol{\theta}}_f = \mathbf{0}$. Therefore, the system is not affected by faults and the detection is complete. If this value is different from zero, then the system is affected by faults and the detection is complete. However, due to model uncertainties noise in the measurements and other disturbances the value of the objective function is not equal to zero, even when the estimated fault vector $\hat{\boldsymbol{\theta}}_f$ coincides with the real fault vector $\boldsymbol{\theta}_f$. Moreover, the fault vector can be different from zero even when the system is not affected by faults. This causes uncertainties in the decision. The introduction of the f_{umbral} in Step 2 of the methodology tries to avoid this disadvantage: The threshold value of the objective function, f_{umbral} in Step 2, should be related with the value of the noise that is affecting the system, and it depends on the system to be diagnosed. In Chap. 4 are shown examples on how to incorporate the prior information about the noise level affecting the system in order to determine f_{umbral}.

The value of f_{umbral} is computed prior to the application of the Step 2 of FD-IPM. It is logic to think that f_{umbral} should depend on the external disturbances affecting the system, e.g., noise. Therefore, deciding the value of f_{umbral} needs some efforts. Therefore, a value of f_{umbral} is a key point for a correct detection of faults in systems described in Situation 1 (Application of Step 2 of the FD-IPM).

If the system satisfies hypotheses H1, H2, and H3, then Step 2 is avoided, and one may go directly to Step 3. The Step 3 of the methodology is very important. At this step, the optimization problem is solved with the help of metaheuristics, and the estimated value for the fault vector is obtained. Based on this estimation, in Step 4 of FD-IPM, the conclusion about the diagnosis of the system is given. As a consequence, the reliability of the diagnosis depends on the performance of the metaheuristics when solving the optimization problem formulated in Step 1 of the FD-IPM. In the solution of the optimization problem in Step 3, the information about the external disturbances or noise affecting the system can be incorporated in the stopping criterion of the metaheuristic methods used (similar as it is incorporated in the f_{umbral} of Step 2, see Chap. 4 for more details). Moreover, the metaheuristics may be initiated considering that $\hat{\boldsymbol{\theta}}_f^{(0)} = \mathbf{0}$ is the initial solution candidate. With this, the Step 2 is incorporated in Step 3.

In Fig. 2.3 it is shown a diagram with the steps of the methodology.

The proposed methodology allows to alleviate the following limitations of the current model based FDI methods:

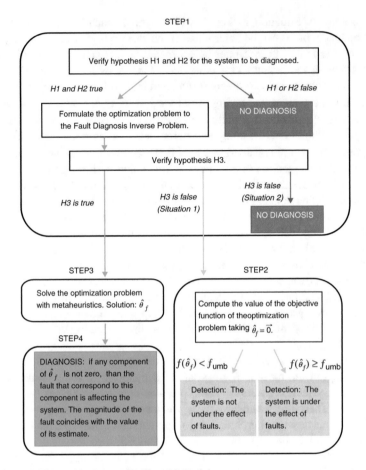

Fig. 2.3 Fault Diagnosis—Inverse Problem Methodology

- The direct fault estimation instead of the model parameters estimation allows to simplify the diagnosis: it does not need to know the relation parameters model/physical parameters of the system/faults.
- It is a general methodology, it does not depend on the model of the system once it incorporates the faults.
- FD-IPM also permits the direct estimation of the faults indistinctly if they take place in actuators, process, or sensors.
- It does not need additional efforts for generating robust residuals: the application of robust metaheuristics allows robust fault estimations. Moreover, metaheuristic control parameters can be handled and tuned in order to improve robustness and sensitivity of the diagnosis.
- The use of metaheuristics and the analysis of the influence of their parameters allow to reduce the computational cost, which is an important issue related to diagnosis time.

2.2 Mathematical Models for the Fault Diagnosis: Inverse Problem Methodology

This section describes the main models, which are used to describe a system, and how they incorporate the fault vector $\boldsymbol{\theta}_f$.

2.2.1 State Space Model

Within the State Space Models, it is possible to distinguish two groups: nonlinear models and linear models (it also includes linearized models).

For nonlinear models, see Eq. (1.1), the fault vector $\boldsymbol{\theta}_f$ is incorporated as:

$$\begin{aligned} \dot{x}(t) &= f\left(x(t), u(t), \boldsymbol{\Theta}, \boldsymbol{\theta}_f\right) \\ y(t) &= g\left(x(t), u(t), \boldsymbol{\Theta}, \boldsymbol{\theta}_f\right) \end{aligned} \tag{2.6}$$

The use of Linear Time Invariant (LTI) models allows obtaining a general structure for incorporating the faults. Let's consider, as in hypothesis H1 (see Sect. 2.1), that each of the m actuators may be affected by only one fault at each time instant: $f_u = (f_{u1}\ f_{u2}\ \cdots\ f_{um})^T$; each of the p output sensors may be affected by only one fault at each time instant: $f_y = (f_{y1}\ f_{y2}\ \cdots\ f_{yp})^T$; and q faults may occur in the process: $f_p = (f_{p1}\ f_{p2}\ \cdots\ f_{pq})^T$. Now the model represented by Eq. (1.2), but taking the faults in consideration, is written as:

$$\begin{aligned} \dot{x}(t) &= Ax(t) + Bu(t) + E_f\boldsymbol{\theta}_f \\ y(t) &= Cx(t) + Du(t) + F_f\boldsymbol{\theta}_f \end{aligned} \tag{2.7}$$

The matrices A, B, C, and D coincide with those from the representation of the system under no faults, see Eq. (1.2). The matrices E_f and F_f are also well known:

$$E_f = (B\ E_p\ \mathbf{0}_{n\times p}) \qquad F_f = (D\ \mathbf{0}_{p\times q}\ I_{p\times p}) \tag{2.8}$$

where the matrix $E_p \in \mathbb{M}_{n\times q}(\mathbb{R})$ represents the influence of the process faults; I is an identity matrix [31]; $E_f \in \mathbb{M}_{n\times(m+q+p)}(\mathbb{R})$; and $F_f \in \mathbb{M}_{p\times(m+q+p)}(\mathbb{R})$.

2.2.2 Models Based on Transfer Function

Let's consider explicitly the faults in a Single Input Single Output (SISO) process, which is represented by a transfer function. Then the model is represented by the following equation:

$$y(s) = G_{yu}(s)u(s) + G_{yf_u}(s)f_u(s) + G_{yf_y}(s)f_y(s) + G_{yf_p}(s)f_p(s) \tag{2.9}$$

where $u(s) \in \mathbb{R}$ represents the output of the system. The transfer function $G_{yu}(s)$ represents the dynamic of the system, and $G_{yf_u}(s)$, $G_{yf_p}(s)$, and G_{yf_y} are the transfer functions that represent the faults f_u, f_p, and f_y, respectively [31].

2.3 Structural Analysis of Fault Diagnosis Inverse Problems: Verifying Hypothesis H3

As described in Sect. 1.3, the main characteristic of Inverse Problems is that they are usually ill-posed problems. This makes them hard to solve due to no guarantee of uniqueness or stability of the solution.

In the case of Fault Diagnosis Inverse Problems, as formulated in Sect. 2.1, the non-uniqueness or non-stability implies that three may exist more than one set of values for $\hat{\theta}_f$ that leads to the same observed behavior of the system. As a consequence, a wrong diagnosis may be obtained.

The problem of detectability and isolability of faults can be used to understand the problem of existence and uniqueness in Fault Diagnosis Inverse Problems. For that reason, and in order to obtain some prior information about the existence and uniqueness of the solution of Fault Diagnosis Inverse Problems, some results related with sensors placement for fault detectability and separability are applied [77, 78].

With the structural analysis of a model of a system, it is possible to decide if the current sensors of the system make possible detecting and isolating the faults that are modeled [77, 78]. It means that it is possible to determine if vector θ_f can be estimated in a unique way.

Detectability of a fault indicates if the effect of the fault in the system can be monitored. It means that it is possible to recognize if the system is under the effect of a fault or not. It could be understood as determining if for the Fault Diagnosis Inverse Problem for a certain system to be diagnosed, the solution exists.

The condition of isolability is related with the separability of a fault from the other faults that may be eventually affecting the system. It means, that the isolability of the faults in a system is needed, in order to complete the diagnosis. This is directly linked with the uniqueness of the inverse problem solution.

Some recent works show that information concerning this topic may be extracted from the structural representation of the model [77, 78]. These results are based on the description of the model as a bipartite graph and its Dulmage-Mendelsohn Decomposition [78].

Let's denote $E = \{e_1, e_2, \ldots e_h\}$, with $h \geq (n + m)$, the set of equations in the model (linear or nonlinear) of a system; and $X = \{x_1, x_2..x_k\}$, with $k \geq n$, the set of variables. Let's also assume that each fault f_l only affects one equation e_{f_l}. Let's construct the biadjacency matrix $\mathcal{M}$ of the bipartite graph $G_b = (E, X)$, which represents the structural information of the model $\mathfrak{M}$ formed by equations in E and variables in X, whose element m_{ij} is equal to one if the variable x_j or its first derivative $\dot{x}_j$ appears in equation e_i, and it is zero in other case. This can be summarized as:

$$m_{ij} = \begin{cases} 1 \text{ if} & x_j \vee \dot{x}_j \in e_i \\ 0 \text{ otherwise} \end{cases} \tag{2.10}$$

where $i = 1, 2, \ldots, h$ and $j = 1, 2, \ldots, k$.

Definition 2.1 A set of equations E is **structurally overdetermined** with respect to a set of variables X if and only if

$$\forall \acute{X} \subseteq X, \acute{X} \neq \phi : \left|\acute{X}\right| < \left|ec_E \acute{X}\right| \tag{2.11}$$

where $|P|$ denotes the cardinality of the set P, and $ec_E \acute{X}$ denotes the set of equations E that contains all the variables in $\acute{X}$ [36].

An alternative definition indicates that it corresponds to the minimal subset in E such that it has the maximum *surplus*. The definition of surplus is presented next.

Definition 2.2 Let's consider a bipartite graph $G_b = (E, X)$ and $\acute{E} \subseteq E$, the **surplus** of $\acute{E}$ is defined as:

$$\varphi(\acute{E}) = \left|\acute{E}\right| - \left|var(\acute{E})\right| \tag{2.12}$$

where $var(P)$ denotes the number of unknown variables that appear in equations of P [36].

Definition 2.2 indicates that the surplus of a subset of equations $\acute{E}$ is the differences between its cardinality and the cardinality of the subset of variables that appear in the equations of $\acute{E}$.

It has been demonstrated that there is a unique overdetermined part of a model, see [36]. It has also been proved that based on the Dulmage-Mendelsohn Decomposition, also known as canonical decomposition, it is possible to obtain such part. The decomposition divides the model into three parts [78]:

- **Structural overdetermined part** $\mathfrak{M}^+ = \mathfrak{M}_\infty$.
- **Determined part** $\mathfrak{M}^- = \bigcup_{i=1}^{n} \mathfrak{M}_i$.
- **Underdetermined part** $\mathfrak{M}^0$.

In terms of the graph theory, the Dulmage-Mendelsohn Decomposition finds a maximal matching in the bipartite graph $G_b = (E, X)$. Let's call E_m and X_m the sets of equations and variables, respectively, which are in the maximal matching. The following definitions are equivalent [36]:

Definition 2.3 The **overdetermined part of a model** $\mathfrak{M}$ is the set of equations such that there is an alternate path in $\mathfrak{M} \backslash E_m$. Its notation is $\mathfrak{M}^+$.

Definition 2.4 The **undetermined part of a model** $\mathfrak{M}$ is the set of equations such that there is an alternate path in $X \backslash X_m$. Its notation is $\mathfrak{M}_0$.

Definition 2.5 The **determined part of a model** $\mathfrak{M}$ is $\mathfrak{M}^- = \mathfrak{M} \backslash (\mathfrak{M}^+ \cup \mathfrak{M}_0)$.

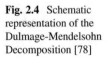

Fig. 2.4 Schematic
representation of the
Dulmage-Mendelsohn
Decomposition [78]

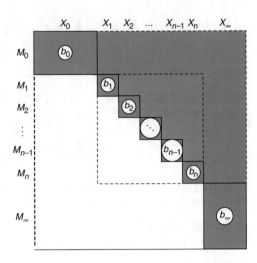

Therefore, the decomposition defines a partition $\mathfrak{M}_0, \mathfrak{M}_1, \ldots \mathfrak{M}_n, \mathfrak{M}_\infty$ in E, which allows to define a partition in X. This allows to establish a partial order among $\mathfrak{M}_i$. Each pair $(\mathfrak{M}_i, X_i)$ is linked with a block b_i.

Definition 2.6 The blocks $b_1, b_2, \ldots b_n$ from the determined part of a model are called ***strongly connected components***.

In Fig. 2.4 is presented a schematic representation of the Dulmage-Mendelsohn Decomposition.

The previously presented results were applied for determining detectability and isolability of faults; as well as the minimum number of sensors required in a system in order to reach certain predefined diagnosis criteria [77].

The overdetermined part of a model indicates that there is a larger number of equations than unknown variables. In other words, this part presents a certain analytical redundancy, which permits the monitoring of the system. Therefore, this part is very useful in FDI and it is also known as monitorable part of the system..

It is clear that a fault cannot be detected if it does not appear in the monitorable part of the system. As a consequence, the following definitions are introduced:

Let $\mathfrak{M}^+$ be the overdetermined part of the model $\mathfrak{M}$. The structural detectability is defined as:

Definition 2.7 The fault f_l is ***structurally detectable*** if and only if $e_{fl} \in \mathfrak{M}^+$ *[77]*.

Definition 2.7 indicates that a fault f_l is structurally detectable, if there exists an observation that is consistent with the fault mode f_l, and inconsistent with the non-fault mode. In other words, the equation e_{fl} which is affected by the fault f_l appears in the monitorable part of the model $\mathfrak{M}^+$. It means that when the fault f_l affects the system, the monitorable part of the system changes. This is known in FDI as a *violation of the monitorable part of the system*. It is concluded that a fault f_l is structurally detectable if f_l can violate the monitorable part of the system.

On the other hand, the definition of structurally separable is [77]:

Definition 2.8 A fault f_l can be *isolated or separated* from another fault f_j in a model $\mathfrak{M}$ if and only if $e_{fl} \in (\mathfrak{M} \backslash e_{fj})^+$ *[77].*

The Dulmage-Mendelsohn Decomposition also gives the order among the strongly connected components, b_i, of $\mathfrak{M}$. This order allows to know which variables must be measured in order to make a certain equation e_h part of $\mathfrak{M}^+$. The following lemma shows this result [77]:

Lemma 2.2 *Let $\mathfrak{M}$ be the determined part of a set of equations; b_i a strongly connected component in $\mathfrak{M}$ with equations $\mathfrak{M}_i$; and $e_h \notin \mathfrak{M}$ an equation that measures the variable in b_i. Then, it is obtained that [77]:*

$$(\mathfrak{M} \cup e_h)^+ = e_h \cup (\cup_{\mathfrak{M}_j \leq \mathfrak{M}_i} \mathfrak{M}_j) \qquad (2.13)$$

See [77], for demonstration of this lemma. Based on this lemma, it is clear that a fault f_j can be detected measuring any of the variables that appear in a bigger block (following the order of the strongly connected components) than the block, which is affected by the fault under study f_j. The isolability or separability can be understood as a particular detectability problem. As a consequence, Lemma 2.2 can also be applied to the isolability analysis.

These results will be applied in order to know if the system under analysis satisfies the requirements for detectability and isolability of its faults when no extra sensors are added. In other words, one wants to know if the fault vector can be correctly estimated when only the information from the current sensors of the system is available. This is the way to verify hypothesis H3 of FD-IPM. It is an alternative existence-uniqueness analysis for the Fault Diagnosis Inverse Problem.

In Refs. [77, 78] it is presented an efficient algorithm and its implementation in Matlab, which allows to make the structural analysis in systems with a very large number of equations. In the study cases considered in this book the Dulmage-Mendelsohn Decomposition is obtained with the function **dmperm** from Matlab®.

2.4 Description of Three Benchmark Problems

This section describes the three benchmark problems that will be used in the numerical experiments in Chap. 4.

The benchmark problems selected are:

- DC Motor: Its model is represented by a transfer function. It is a SISO system. It can be affected by three faults: actuator fault, process fault, and sensor fault.
- Inverted Pendulum System (IPS): Its model is represented by an LTI model in state variables. It has one input and two outputs. The system can be affected by three faults: actuator fault and two sensors faults.

- Two Tanks System: Its model is nonlinear in state variables. It has two inputs and two outputs. The system can be affected by two faults: two sensor faults.

For each benchmark is made a general description of the system and corresponding faults, as well as a presentation and description of the mathematical model.

2.4.1 Benchmark 1: DC Motor

The DC Motor control system DR300 has been widely used for studying and testing new methods of FDI, due to its similitude with high speed industrial control systems [31].

The system is formed by a permanent magnet, which is coupled to a DC generator. The main function of this generator is to simulate the effect of a fault that results when a load torque is applied to the axis of the motor. The speed is measured by a tachometer, which feeds the signal to a Proportional-Integral (PI) Speed Controller. In Fig. 2.5 is shown the block diagram of the DC Motor control system AMIRA DR300.

The voltage U_T is proportional to the rotational speed of the motor axis ω. U_T is compared with U_{ref} in order to use the error for computing the control signal U_C for the PI speed controller. The AMIRA DR300 system also includes an internal control loop for the armature current I_A. The controller computes the motor armature voltage U_A as a function of the reference, that is obtained by means of the gain K_1 and the output I_A. The units are volts (V) for U_T, U_{ref}, U_C and U_A; amperes (A) for I_A; rpm for ω; and A/V for K_1.

The technical parameters considered in the DC Motor benchmark problem are shown in Table 2.1.

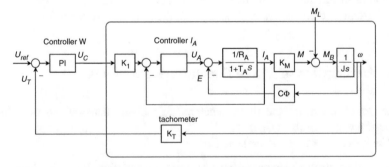

Fig. 2.5 Block diagram of the DC Motor control system AMIRA DR300 [31]

Table 2.1 Values of the parameters for the laboratory DC Motor DR300 [31]

Element	Parameter	Value
Motor	Nominal tension, U_n	24 V
	Nominal speed, V_n	3000 rpm
	Nominal current, I_n	2 A
	Nominal power, P_n	30 W
	Nominal torque, M_n	0.096 Nm
	Starting torque, M_{arr}	0.4 Nm
	Armature resistance, R_a	2.6 Ohm
	Armature inductance, L_a	0.003 H
	Motor's constant, K_M	6 Nm/A
	Total inertia, J	0.06 Nm/A
	Tension constant, $C\phi$	6.27 mV/rpm
	Gain, K_1	0.4 A/V
Tachometer	Gain, K_T	5 mV/rpm
	Maximum speed	5000 rpm
	Charge impedance	2500 Ohm
Controller	Proportional gain, K_p	1.96
PI of ω	Integral gain, K_i	1.6

2.4.1.1   Mathematical Model

For this study, the internal current loop, the DC Motor, and the tachometer are considered as a single block. The block diagram of the closed loop is formed by the process and the PI controller.

The dynamics of the control system in the open loop is described in the frequency domain by:

$$U_T(s) = G_{yu}(s)U_C(s) + G_d(s)M_L(s) \tag{2.14}$$

$$G_{yu}(s) = \frac{8.75}{(1 + 1.225s)(1 + 0.03s)(1 + 0.005s)} \tag{2.15}$$

$$G_d(s) = -\frac{31.07}{s(1 + 0.005s)} \tag{2.16}$$

where U_T is the controlled variable; U_C is the control signal, M_L is the torque applied in the axis of the motor, and $G_d(s)$ its transfer function. $G_{yu}(s)$ represents the dynamics of the process in open loop [31].

The dynamics of the control system in the open loop is represented in Fig. 2.6. As it is shown, the PI controller does not take the feedback of the output in the open loop.

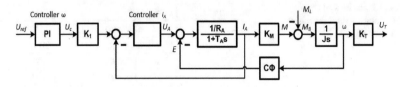

Fig. 2.6 Block diagram of AMIRA DR300 DC Motor control system in open loop [31]

The transfer function of the PI speed controller is:

$$G_c(s) = \frac{U_C(s)}{E(s)} = 1.96 + \frac{1.6}{s} \tag{2.17}$$

$$E(s) = U_{ref}(s) - U_T(s) \tag{2.18}$$

where $E(s)$ denotes the error signal.

2.4.1.2 Modeling the Faults

The system may be affected by three additive faults $\boldsymbol{\theta}_f = (f_u \; f_p \; f_y)^T$:

- Fault 1: f_u represents a fault in actuator. It is modeled as a deviation of the control signal.
- Fault 2: f_y represents a fault in the sensor that measures the motor speed.
- Fault 3: $f_p = M_L$ represents a fault in the process itself due to a load torque, which acts on the axis of the motor.

The following equation describes the system response under the effect of faults by means of a closed loop transfer function:

$$U_T(s) = G_{yu}(s)U_{ref}(s) + G_{yf_u}(s)f_u(s) + G_{yf_y}(s)f_y(s) + G_d(s)f_p(s) \tag{2.19}$$

It is assumed that the faults are time invariant once they appear. The restrictions for the components of the fault vector are:

$$
\begin{aligned}
f_u, f_y \in \mathbb{R} : -1V \leq f_u, f_y \leq 1\,V \\
f_p \in \mathbb{R} : -1\,\text{Nm} \leq f_p \leq 1\,\text{Nm}
\end{aligned}
\tag{2.20}
$$

2.4.1.3 Simulations of the Benchmark DC Motor

All simulations of the speed control system for the closed loop presented in this book were performed with Matlab®. In all test cases, it was considered that the system was affected by noise. The reference speed is 3000 rpm, which corresponds to 15 V.

Fig. 2.7 Inverted Pendulum
System (IPS) [97]

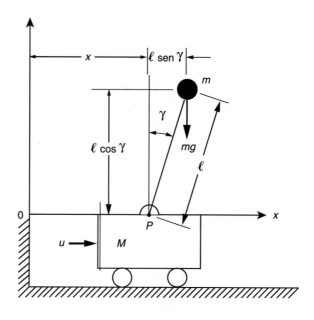

2.4.2   Benchmark 2: Inverted Pendulum System (IPS)

This system is also considered as a benchmark for control and diagnosis [31]. It is
formed by an inverted pendulum with mass m and length l mounted on a motor-
driven car with mass m_c. The objective is to keep the beam aligned with the vertical
position. Here, it has been considered only a two-dimensional problem, where the
pendulum only moves in the plane of the paper, see Fig. 2.7.

2.4.2.1   Mathematical Model

The mathematical model of the IPS has been widely studied [31, 97]. The system
is described by a state-space representation of a linear time invariant system (LTI
model). The state vector is $x = (\gamma \; \dot{\gamma} \; x \; \dot{x})^T$, where γ and $\dot{\gamma}$ are the angle of the
pendulum with respect to the vertical position and the angular velocity, respectively;
and x and $\dot{x}$ represent the car position and velocity, respectively. The output vector
of the system is $y = (\gamma \; x)^t$. The input $u = F$ is the control force applied to the car.
The LTI model in state variables is:

$$\dot{x}(t) = Ax(t) + Bu(t)$$
$$y(t) = Cx(t) \tag{2.21}$$

being matrices A, B, and C given by:

$$A = \begin{pmatrix} 0 & 1 & 0 & 0 \\ \frac{m+m_c}{m_c l}g & 0 & 0 & 0 \\ 0 & 0 & 0 & 1 \\ -\frac{m}{m_c}g & 0 & 0 & 0 \end{pmatrix} \quad B = \begin{pmatrix} 0 \\ -\frac{1}{m_c l} \\ 0 \\ -\frac{1}{m_c} \end{pmatrix} \tag{2.22}$$

$$C = \begin{pmatrix} 1 & 0 & 0 & 0 \\ 0 & 0 & 1 & 0 \end{pmatrix} \tag{2.23}$$

where g is the gravity acceleration, and the other mathematical symbols have already been defined.

2.4.2.2   Modeling the Faults

The system may be affected by three additive faults:

- Fault 1: f_u represents an actuator fault that causes undesired movement of the car.
- Fault 2: $f_{y(1)}$ represents a fault in the sensor that measures γ.
- Fault 3: $f_{y(2)}$ represents a fault in the sensor that measures x.

The fault vector is $\boldsymbol{\theta}_f = \begin{pmatrix} f_u & f_{y(1)} & f_{y(2)} \end{pmatrix}^T$, and the system with the faults is described by the model:

$$\begin{aligned} \dot{\boldsymbol{x}}(t) &= \boldsymbol{A}\boldsymbol{x}(t) + \boldsymbol{B}u(t) + \boldsymbol{E}_f\boldsymbol{\theta}_f \\ \boldsymbol{y}(t) &= \boldsymbol{C}\boldsymbol{x}(t) + \boldsymbol{F}_f\boldsymbol{\theta}_f \end{aligned} \tag{2.24}$$

being $\boldsymbol{E}_f \in \mathbb{M}_{4\times3}(\mathbb{R})$. In the present model $\boldsymbol{D} = \boldsymbol{0}_{2\times1}$, and therefore $\boldsymbol{F}_f \in \mathbb{M}_{2\times3}(\mathbb{R})$:

$$\boldsymbol{E}_f = (\boldsymbol{B} \; \boldsymbol{0}_{4\times2}), \quad \boldsymbol{F}_f = (\boldsymbol{0}_{2\times1} \; \boldsymbol{I}_{2\times2}) \tag{2.25}$$

It is assumed that the faults keep constant values once they appear.

Considering the nature of faults and the properties of the IPS under study, see Table 2.2, the elements of $\boldsymbol{\theta}_f$ have the following restrictions:

$$f_u \in \mathbb{R} : \; -0.5 \text{ N} \leq f_u \leq 0.5 \text{ N} \tag{2.26}$$

$$f_{y(1)} \in \mathbb{R} : \; 0 \text{ rad} \leq f_{y(1)} \leq 0.01 \text{ rad} \tag{2.27}$$

$$f_{y(2)} \in \mathbb{R} : \; 0 \text{ m} \leq f_{y(2)} \leq 0.02 \text{ m} \tag{2.28}$$

Table 2.2 Parameter values for the IPS under analysis

Parameter	Value
Mass of the car, m_c	2 kg
Mass of the pendulum, m	0.1 kg
Length of the pendulum, l	0.5 m

2.4.2.3 Simulations of the Benchmark IPS

Considering the system with the reported characteristics in Ref. [97], which leads to the parameter values shown in Table 2.2, the following matrices are obtained:

$$A = \begin{pmatrix} 0 & 1 & 0 & 0 \\ 20.601 & 0 & 0 & 0 \\ 0 & 0 & 0 & 1 \\ -0.4905 & 0 & 0 & 0 \end{pmatrix} \quad B = \begin{pmatrix} 0 \\ -1 \\ 0 \\ 0.5 \end{pmatrix} \tag{2.29}$$

$$C = \begin{pmatrix} 1 & 0 & 0 & 0 \\ 0 & 0 & 1 & 0 \end{pmatrix} \tag{2.30}$$

Using the parameter values from Table 2.2, the following numerical values for the matrices E_f and F_f are obtained:

$$E_f = \begin{pmatrix} 0 & 0 & 0 \\ -1 & 0 & 0 \\ 0 & 0 & 0 \\ 0.5 & 0 & 0 \end{pmatrix}, \quad F_f = \begin{pmatrix} 0 & 1 & 0 \\ 0 & 0 & 1 \end{pmatrix} \tag{2.31}$$

The behavior of the system was simulated both free of faults and under different faulty situations. The direct problem in Eq. (2.24) was numerically solved with the fourth order Runge-Kutta method. All the implementations were made in Matlab®.

2.4.3 Benchmark 3: Two Tanks System

The Two Tanks System was adopted as a benchmark for control and diagnosis starting in the 90s of the last century [31, 86], see Fig. 2.8. It is a simplified version from another benchmark, the Three Tanks System [86]. The mathematical symbols shown in Fig. 2.8 are described in the next section.

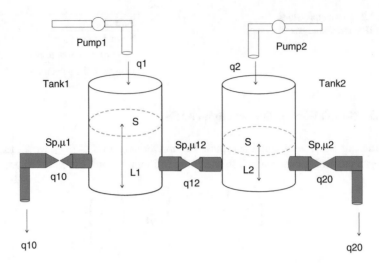

Fig. 2.8 Two Tanks System

2.4.3.1 Mathematical Model

The system is formed by two tanks of liquid. Each tank is filled with an independent pump. Both pumps are similar, and both tanks have the same value for the cross section $S_1 = S_2 = S$. In order to keep the liquid level at certain nominal values, $\tilde{L}_1$ and $\tilde{L}_2$, in tanks 1 and 2, respectively, the pumps feed the tanks with the flow rates q_1 and q_2, respectively. The tanks are interconnected and drained with lower pipes, as shown in Fig. 2.8. All pipes have the same cross section S_p.

The liquid levels L_1 and L_2, at tanks 1 and 2, respectively, are the outputs of the system, $\boldsymbol{y} = (L_1 \ L_2)^T$. They are measured with sensors. The control variables are q_1 and q_2, $\boldsymbol{u} = (q_1 \ q_2)^T$.

The system model is derived from the application of the Conservation of Mass Principle and considering that the flow obeys Torricelli's law:

$$\begin{cases} \dot{L}_1 = \frac{q_1}{S} - \frac{C_1}{S}\sqrt{L_1} - \frac{C_{12}}{S}\sqrt{|L_1 - L_2|}\,\text{sign}\,(L_1 - L_2) \\ \dot{L}_2 = \frac{q_2}{S} - \frac{C_2}{S}\sqrt{L_2} + \frac{C_{12}}{S}\sqrt{|L_1 - L_2|}\,\text{sign}\,(L_1 - L_2) \\ y_1 = L_1 \\ y_2 = L_2 \end{cases} \tag{2.32}$$

where $C_1 = \mu_1 S_p \sqrt{2g}$, $C_2 = \mu_2 S_p \sqrt{2g}$, and $C_{12} = \mu_{12} S_p \sqrt{2g}$. The second term on the right-hand side of the first two equations represents the outflow in each tanks, q_{10} and q_{20}; and the third term represents the value of the flow rate between the two tanks, q_{12}. Here, μ_1, μ_2, and μ_{12} are related to the resistance to the flow in the lower pipes.

Table 2.3 Parameter values used in the simulation of the Two Tanks System

Parameter	Value
C_1, C_2, C_{12}	$0.3028 \, \text{m}^{5/2}/\text{s}$
$S_1 = S_2 = S$	$2.54 \, \text{m}^2$
S_p	$0.1 \, \text{m}^2$
Gravity acceleration, g	$9.8 \, \text{m/s}^2$

2.4.3.2   Modelling the Faults

The system can be affected by two process faults:

- Fault 1: $f_{p(1)}$ is the output flow rate value of an undesired leakage in *Tank 1*.
- Fault 2: $f_{p(2)}$ is the output flow rate value of an undesired leakage in *Tank 2*.

The fault vector is $\boldsymbol{\theta}_f = \left(f_{p(1)} \; f_{p(2)} \right)^T$. The nonlinear model in state variables of the Two Tanks System with the incorporation of the faults is:

$$\begin{cases} \dot{L}_1 = \frac{q_1}{S} - \frac{C_1}{S}\sqrt{L_1} - \frac{C_{12}}{S}\sqrt{|L_1 - L_2|}\text{sign}\,(L_1 - L_2) - \frac{f_{p(1)}}{S} \\ \dot{L}_2 = \frac{q_2}{S} - \frac{C_2}{S}\sqrt{L_2} + \frac{C_{12}}{S}\sqrt{|L_1 - L_2|}\text{sign}\,(L_1 - L_2) - \frac{f_{p(2)}}{S} \\ y_1 = L_1 \\ y_2 = L_2 \end{cases} \qquad (2.33)$$

As a first approximation, it is assumed that the leaks in both tanks do not change with time. It is also assumed that the leaks do not allow a flow rate bigger than $1 \, \text{m}^3/\text{s}$. This restriction can be modeled as:

$$f_{p(1)}, f_{p(2)} \in \mathbb{R} : \; 0 \le f_{p(1)}, f_{p(2)} \le 1 \, \text{m}^3/\text{s} \qquad (2.34)$$

2.4.3.3   Simulations of the Benchmark Two Tanks System

The parameter values used in the simulation of the Two Tanks System are presented in Table 2.3. A PID controller has been designed with parameters $\boldsymbol{K_p} = (12 \; 14)$; $\boldsymbol{K_i} = (1.15 \; 0.3)$ and $\boldsymbol{K_d} = (1.0 \; 1.5)$. The nominal values for the liquid level in tanks are $\tilde{L}_1 = 4.0 \, \text{m}$ and $\tilde{L}_2 = 3.0 \, \text{m}$,

The implementations were made in Matlab®. The direct problem was solved using the fourth order Runge Kutta method.

2.5   Application of the First Step of the Fault Diagnosis: Inverse Problem Methodology

This section presents the application of the first step of the Fault Diagnosis—Inverse Problem Methodology (FDIPM) to each benchmark problem. In this first step, it is formulated the FDI as an Inverse Problem for each benchmark, followed by the corresponding optimization problem, see Eq. (2.3), to be solved in step number three of FD-IPM. The first step contains also the verification of hypothesis H1–H3, see Sect. 2.1.

The optimization problems formulated for the three benchmarks considered are solved in Chap. 4, with the use of metaheuristics.

2.5.1 Benchmark Problem: DC Motor

The model that describes the DC Motor, which can be affected by three faults, was presented in Sect. 2.4.1, see Eq. (2.19). The model and its incorporation of the three faults satisfy hypothesis H1. It is also assumed that the faults are time invariant once they appear. As a consequence, the system also satisfies hypothesis H2.

The optimization problem that corresponds to the Fault Diagnosis Inverse Problem for this benchmark is derived from Eq.(2.1), since there is only one output, and considering also restrictions to the fault values given by Eq. (2.20), it is written:

$$\min \ f(\hat{\boldsymbol{\theta}}_f) = \sum_{t=1}^{I} \left[U_{T\,t}(\boldsymbol{\theta}_f) - \hat{U}_{T\,t}(\hat{\boldsymbol{\theta}}_f) \right]^2$$

$$\text{s.t.} \qquad\qquad\qquad (2.35)$$

$$\hat{f}_u, \hat{f}_y \in \mathbb{R} : -1\,\mathrm{V} \le \hat{f}_u, \hat{f}_y \le 1\,\mathrm{V}$$

$$\hat{f}_p \in \mathbb{R} : -1\,\mathrm{Nm} \le \hat{f}_p \le 1\,\mathrm{Nm}$$

where $U_{T\,t}(\boldsymbol{\theta}_f)$ is measured with a sensor that can present a fault with value f_y; $\hat{U}_{T\,t}(\hat{\boldsymbol{\theta}}_f)$ is the calculated value of the motor speed obtained with Eq. (2.19) and the Inverse Laplace Transform.

In order to diagnose the DC Motor based on the FD-IPM, the optimization problem described by Eq. (2.36) is solved in FD-IPM Step 3 with the help of metaheuristics.

2.5.2 Benchmark Problem: Inverted Pendulum System

The model that describes the IPS, which can be affected by three faults, was presented in Sect. 2.4.2, see Eq. (2.24). The model and its incorporation of the three faults satisfy hypothesis H1. It is also assumed that the faults keep constant values once they appear. As a consequence, the system satisfies hypothesis H2.

Considering the optimization problem described by Eq. (2.3) and the restrictions represented by Eqs. (2.26)–(2.28), it is obtained the following optimization problem, after applying the FD-IPM Step 1:

$$\min \ f(\hat{\boldsymbol{\theta}}_f) = \left\| \sum_{t=1}^{I} \left[\boldsymbol{y}_t(\boldsymbol{\theta}_f) - \hat{\boldsymbol{y}}_t(\hat{\boldsymbol{\theta}}_f) \right]^2 \right\|_\infty$$

$$\text{s.t.} \qquad (2.36)$$

$$\hat{f}_u \in \mathbb{R} : -0.5\,\mathrm{N} \leq \hat{f}_u \leq 0.5\,\mathrm{N}$$

$$\hat{f}_{y(1)} \in \mathbb{R} : 0\,\mathrm{rad} \leq \hat{f}_{y(1)} \leq 0.01\,\mathrm{rad}$$

$$\hat{f}_{y(2)} \in \mathbb{R} : 0\,\mathrm{m} \leq \hat{f}_{y(2)} \leq 0.02\,\mathrm{m}$$

being $\boldsymbol{y}_t = (y(t)\,x(t))^T$ the values of the output at the instant of time t measured with sensors; and $\hat{\boldsymbol{y}}_t = (\hat{y}(t)\,\hat{x}(t))^T$ are the calculated values obtained with the model given by Eq. (2.24) at the instant of time t.

In order to diagnose the IPS based on the FD-IPM, the optimization problem described by Eq. (2.37) is solved in the FD-IPM Step 3 with the help of metaheuristics.

2.5.2.1   Structural Analysis: Verifying Hypothesis H3

Next will be presented the verification of hypothesis H3 for the Fault Diagnosis Inverse Problem of the IPS, which was formulated as an optimization problem in Eq. (2.37). For that, the structural analysis of the direct model of the IPS, see Eq. (2.24), need to be made.

In order to show clearly the results presented in Sect. 2.3, the Structural Analysis of the IPS is made in a constructive way, i.e. the faults and their equations are added step by step until the Structural Analysis of the IPS model described in Sect. 2.4.2 is obtained, see Eq. (2.24).

Let's denote as follows the equations that represent the model of the IPS under study:

$$e_1 : \dot{x}_1 = x_2 \qquad (2.37)$$

$$e_2 : \dot{x}_2 = 20.601\,x_1 - u - f_u \qquad (2.38)$$

$$e_3 : \dot{x}_3 = x_4 \qquad (2.39)$$

$$e_4 : \dot{x}_4 = -0.4905\,x_1 + 0.5\,u + 0.5\,f_u \qquad (2.40)$$

$$e_6 : y_2 = x_3 + f_{y(2)} \qquad (2.41)$$

$$e_7 : y_1 = x_4 + f_{y(1)} \qquad (2.42)$$

Let's start considering the model of the IPS when no sensors are present, i.e. removing equations e_6 and e_7 from the model given by Eqs. (2.37)–(2.42). Therefore only an actuator fault f_u is affecting the system. The model of the system in this case is formed by four equations e_1, e_2, e_3, and e_4:

$$e_1 : \dot{x}_1 = x_2 \tag{2.43}$$

$$e_2 : \dot{x}_2 = 20.601 \, x_1 - u - f_u \tag{2.44}$$

$$e_3 : \dot{x}_3 = x_4 \tag{2.45}$$

$$e_4 : \dot{x}_4 = -0.4905 \, x_1 + 0.5 \, u + 0.5 f_u \tag{2.46}$$

In model given by Eqs. (2.43)–(2.46), the fault f_u affects two equations:

$$e_{f_u} = e_2, \, e_4 \tag{2.47}$$

With the aim of satisfying the requirement that a fault affects only one equation, the new artificial variable x_5 and the new equation $e_5 : x_5 = f_u$ are introduced to the model given by Eqs. (2.43)–(2.46). The equations e_2 and e_4 from the model adopted now the form:

$$e_2 : \dot{x}_2 = 20.601 \, x_1 - u - x_5 \tag{2.48}$$

$$e_4 : \dot{x}_4 = -0.4905 \, x_1 + 0.5 \, u + 0.5 x_5 \tag{2.49}$$

The model with the introduction of the artificial variable x_5 and the equation e_5 is described as:

$$e_1 : \dot{x}_1 = x_2 \tag{2.50}$$

$$e_2 : \dot{x}_2 = 20.601 \, x_1 - u - x_5 \tag{2.51}$$

$$e_3 : \dot{x}_3 = x_4 \tag{2.52}$$

$$e_4 : \dot{x}_4 = -0.4905 \, x_1 + 0.5 \, u + 0.5 x_5 \tag{2.53}$$

$$e_5 : x_5 = f_u \tag{2.54}$$

Let's denote by $\mathfrak{M}$ the model given by Eqs. (2.50)–(2.54). In this model the fault f_u only affects equation e_5, i.e. $e_{fu} = e_5$. Considering the bipartite graph of model $\mathfrak{M}$: $G_b = (E, X)$, being $E = \{e_1, \ldots, e_5\}$ and $X = \{x_1, \ldots, x_5\}$, then its biadjacency matrix $\mathcal{M}_{\text{IPS}(1)}$ has the form:

$$
\mathcal{M}_{\text{IPS}(1)} = \begin{array}{c} \\ e_1 \\ e_2 \\ e_3 \\ e_4 \\ e_5 \end{array}
\begin{array}{ccccc}
x_1 & x_2 & x_3 & x_4 & x_5 \\
\left(\begin{array}{ccccc}
1 & 1 & 0 & 0 & 0 \\
1 & 1 & 0 & 0 & 1 \\
0 & 0 & 1 & 1 & 0 \\
1 & 0 & 0 & 1 & 1 \\
0 & 0 & 0 & 0 & 1
\end{array} \right)
\end{array} \tag{2.55}
$$

From this biadjacency matrix, $\mathcal{M}_{\text{IPS}(1)}$ in Eq. (2.55), the Dulmage-Mendelsohn Decomposition is obtained with the help of the function *dmperm* from Matlab®, yielding the matrix in Eq. (2.56). It can be concluded that $\mathfrak{M}^+ = \mathfrak{M}_0 = [\Phi]$ and $\mathfrak{M}^- = \cup_{i=1}^n \mathfrak{M}_i = \mathfrak{M}$. Therefore, $e_{f_u} \notin \mathfrak{M}^+$, and the actuator fault f_u cannot be detected without sensors, which is a logical result.

$$
\mathcal{DMD}_{\text{IPS}(1)} = \begin{array}{c} e_3 \\ e_4 \\ e_1 \\ e_2 \\ e_5 \end{array} \begin{pmatrix} \overset{x_3}{1} & \overset{x_4}{1} & \overset{x_1}{0} & \overset{x_2}{0} & \overset{x_5}{0} \\ 0 & 1 & 1 & 0 & 0 \\ 0 & 0 & 1 & 1 & 0 \\ 0 & 0 & 1 & 1 & 1 \\ 0 & 0 & 0 & 0 & 1 \end{pmatrix} \tag{2.56}
$$

In this case, the order for the strongly connected components $b_1 = (x_3, e_3)$; $b_2 = (x_4, e_4)$; $b_3 = (x_1, x_2, e_1, e_2)$; $b_4 = (x_5, e_5)$ is

$$
b_1 > b_2 > b_3 > b_4 \tag{2.57}
$$

As a consequence of the strongly connected components order and Lemma 2.2 from Sect. 2.3, it is concluded that measuring any of the variables that appear in blocks b_1, b_2, b_3, b_4 permits equation $e_{f_u} = e_5$ to be part of $\mathfrak{M}^+$. It means that incorporating a new equation $e_6 : \ y = x_i$ with $i = 1, 2, 3, 4$, or 5, it is obtained that $e_{f_u} = e_5$ is part of $(\mathfrak{M} \cup e_6)^+$. As a consequence, the incorporation of this equation e_6 makes the fault e_{f_u} structurally detectable, see Definition 2.7 in Sect. 2.3.

For the new equation e_6, it is chosen the form: $e_6 : \ y_2 = x_3$. Its means that it is added a sensor for measuring variable x_3 (position of the car), which corresponds to block b_1. This equation e_6 differs from e_6 in the original model of the IPS under study, see model given by Eqs. (2.37)–(2.42), in the term $f_{y(2)}$ that appears in e_6 from the original model of the IPS. It means that in the original model, the sensor of variable x_3 is affected by a fault which is denoted as $f_{y(2)}$. In order to reach the conditions of the sensor of x_3 in the original problem, let's assume that the new added sensor in the new equation e_6 can be affected by a fault $f_{y(2)}$. Now, equation e_6 has the same form as in the original model $e_6 : \ y_2 = x_3 + f_{y(2)}$. Moreover, the fault $f_{y(2)}$ affects only one equation: $e_{f_{y(2)}} = e_6$. Let's note that the fault $f_{y(2)}$ is detectable with the current sensor because $e_6 = e_{f_{y(2)}} \in (\mathfrak{M} \cup e_6)^+$. This is a consequence of the strongly connected components order of the Dulmage-Mendelsohn Decomposition represented in matrix $\mathcal{DMD}_{\text{IPS}(1)}$, see Eq. (2.56), and Lemma 2.2 from Sect. 2.3: adding equation e_6 to the model with Dulmage-Mendelsohn Decomposition represented in matrix $\mathcal{DMD}_{\text{IPS}(1)}$ makes that all the equations in $\mathcal{DMD}_{\text{IPS}(1)}$ and equation e_6 will be part of the overdetermined part of the new model given by equations $e_1 - e_6$ because e_6 is the equation of a sensor that measures variable x_3, which belongs to block b_1 of the Dulmage-Mendelsohn Decomposition of the model formed by equations $e_1 - e_5$ in matrix $\mathcal{DMD}_{\text{IPS}(1)}$.

At this point, an important question is related to the separability of the faults f_u and $f_{y(2)}$. Based on the definition of faults structurally separable, see Definition 2.8 in Sect. 2.3, f_u will be separated from $f_{y(2)}$ if $e_{f_u} \in ((\mathfrak{M} \cup e_6) \backslash e_{f_{y(2)}})^+$. The model $(\mathfrak{M} \cup e_6) \backslash e_{fy(2)}$ is formed by equations $e_1 - e_5$ which coincide with the equations of model $\mathfrak{M}$. As a consequence, its biadjacency matrix coincides with $\mathcal{M}_{\mathrm{IPS}(1)}$, see Eq. (2.55). Therefore, its Dulmage-Mendelsohn Decomposition and order for the strongly connected components coincide with the presented in Eqs. (2.56) and (2.57), respectively. As a consequence of Lemma 2.2, it is necessary to add another sensor for obtaining that f_u is separable from $f_{y(2)}$. It is decided to add a sensor for measuring variable x_1, i.e. the angle of the pendulum with respect to the vertical position, which forms part of the block b_3 and corresponds to the other sensor that is present in the original model of the IPS. Therefore, there is added a new equation $e_7 : y_1 = x_1$ and:

$$e_{fu} \in (\mathfrak{M} \cup e_7)^+ = \{e_7, e_1, e_2, e_5\} \tag{2.58}$$

being represented the Dulmage-Mendelsohn Decomposition of $\mathfrak{M} \cup e_7$ in the following matrix:

$$\mathcal{DMD}_{\mathrm{IPS}(2)} = \begin{array}{c} \\ e_3 \\ e_4 \\ e_7 \\ e_1 \\ e_2 \\ e_5 \end{array} \begin{array}{ccccc} x_3 & x_4 & x_1 & x_2 & x_5 \\ \begin{pmatrix} 1 & 1 & 0 & 0 & 0 \\ 0 & 1 & 1 & 0 & 0 \\ 0 & 0 & 1 & 0 & 0 \\ 0 & 0 & 1 & 1 & 0 \\ 0 & 0 & 1 & 1 & 1 \\ 0 & 0 & 0 & 0 & 1 \end{pmatrix} \end{array} \tag{2.59}$$

It is observed that with the two added sensors, it is possible to achieve detectability and separability of faults f_u and $f_{y(2)}$. It means that the Fault Diagnosis Inverse Problem when considering only these two faults f_u and $f_{y(2)}$, and the measurements of the two added sensors, but the sensor of x_1 cannot be affected by faults, can be solved. For this case, which is called Case 1, the IPS can be diagnosed with the FD-IPM.

In a similar way, the Fault Diagnosis Inverse Problem by considering only the two faults f_u and $f_{y(1)}$, the measurements of the two considered sensors and with the condition that the sensor x_3 cannot be affected by faults, can be solved. For this case, which is called Case 2, the IPS can be diagnosed with the FD-IPM.

Case 3 represents the IPS with no actuator fault, i.e. it can only be affected by the two sensor faults $f_{y(1)}$ and $f_{y(2)}$. In this case, the artificial variable and equation x_5 and e_5, respectively, do not need to be introduced. Let's denote by $\mathfrak{M}_{case3}$ the model of the system in this case. Its biadjacency matrix $\mathcal{M}_{\mathrm{IPS}(2)}$ has the form:

$$\mathcal{M}_{\mathrm{IPS}(2)} = \begin{array}{c} \\ e_1 \\ e_2 \\ e_3 \\ e_4 \\ e_6 \\ e_7 \end{array} \begin{array}{cccc} x_1 & x_2 & x_3 & x_4 \\ \begin{pmatrix} 1 & 1 & 0 & 0 \\ 1 & 1 & 0 & 0 \\ 0 & 0 & 1 & 1 \\ 1 & 0 & 0 & 1 \\ 1 & 0 & 0 & 0 \\ 0 & 0 & 1 & 0 \end{pmatrix} \end{array} \tag{2.60}$$

The Dulmage-Mendelsohn Decomposition of the model of Case 3, which was represented by the biadjacency matrix $\mathcal{M}_{IPS(2)}$ in Eq.(2.60), is:

$$
\mathcal{DMD}_{IPS(3)} = \begin{array}{c} \\ e_6 \\ e_3 \\ e_4 \\ e_7 \\ e_1 \\ e_2 \end{array} \begin{array}{cccc} x_3 & x_4 & x_1 & x_2 \\ \left(\begin{array}{cccc} 1 & 0 & 0 & 0 \\ 1 & 1 & 0 & 0 \\ 0 & 1 & 1 & 0 \\ 0 & 0 & 1 & 0 \\ 0 & 0 & 1 & 1 \\ 0 & 0 & 1 & 1 \end{array} \right) \end{array} \qquad (2.61)
$$

The matrix $\mathcal{DMD}_{IPS(3)}$ in Eq. (2.61) shows that the overdetermined part of the model $\mathfrak{M}_{case3}$ coincides with the complete model, i.e. $\mathfrak{M}^+_{case3} = \mathfrak{M}_{case3}$. As a consequence, both sensor faults $f_{y(1)}$ and $f_{y(2)}$ are detectable. For the separability analysis, from the Definition 2.8 in Sect. 2.3, $f_{y(2)}$ will be separated from $f_{y(1)}$ if $e_{fy(2)} \in (\mathfrak{M}_{case3} \backslash e_{fy(1)})^+$. The Dulmage-Mendelsohn Decomposition of $\mathfrak{M}_{case3} \backslash e_{fy(1)}$ is:

$$
\mathcal{DMD}_{IPS(4)} = \begin{array}{c} \\ e_6 \\ e_3 \\ e_4 \\ e_1 \\ e_2 \end{array} \begin{array}{cccc} x_3 & x_4 & x_1 & x_2 \\ \left(\begin{array}{cccc} 1 & 0 & 0 & 0 \\ 1 & 1 & 0 & 0 \\ 0 & 1 & 1 & 0 \\ 0 & 0 & 1 & 1 \\ 0 & 0 & 1 & 1 \end{array} \right) \end{array} \qquad (2.62)
$$

The matrix $\mathcal{DMD}_{IPS(4)}$ shows that $(\mathfrak{M}_{case3} \backslash e_{fy(1)})^+ = \mathfrak{M}_{case3} \backslash e_{fy(1)}$. As a consequence, $e_{fy(1)}$ which is e_6 is part of the overdetermined part of $\mathfrak{M}_{case3} \backslash e_{fy(1)}$. It is concluded that for Case 3, the IPS can be diagnosed with FD-IPM.

It is necessary to emphasize that for the IPS benchmark problem under analysis, both sensors and the actuator may have faults. Let's call Case 4 the description of the IPS at its original form, i.e. the three faults can affect the system. In this case, the last added sensor $y_1 = x_1$ is also affected by a fault $f_{y(1)}$, being represented by the following equation:

$$
e_{fy(1)} = e_7 : y_1 = x_1 + f_{y(1)} \qquad (2.63)
$$

The new fault $f_{y(1)}$ is detectable because $e_{fy(1)} \in (\mathfrak{M} \cup e_6 \cup e_7)^+$. Therefore, in the Case 4 all the faults can be detected. It means FD-IPM Steps 1 and 2 can be applied. But regarding separability, the following analysis must be done:

- Analysis 1: f_u separable from $f_{y(2)}$.

   It is necessary to analyze if $e_{fu} \in (\mathfrak{M} \cup e_7)^+$. The Dulmage-Mendelsohn Decomposition of $\mathfrak{M} \cup e_7$ is:

$$\mathcal{DMD}_{\text{IPS(5)}} = \begin{array}{c} \\ e_6 \\ e_3 \\ e_4 \\ e_1 \\ e_2 \\ e_5 \end{array} \begin{array}{ccccc} x_3 & x_4 & x_1 & x_2 & x_5 \\ \left(\begin{array}{ccccc} 1 & 0 & 0 & 0 & 0 \\ 1 & 1 & 0 & 0 & 0 \\ 0 & 1 & 1 & 0 & 0 \\ 0 & 0 & 1 & 1 & 0 \\ 0 & 0 & 1 & 1 & 1 \\ 0 & 0 & 0 & 0 & 1 \end{array} \right) \end{array} \tag{2.64}$$

Matrix $\mathcal{DMD}_{\text{IPS(5)}}$ shows that $(\mathfrak{M} \cup e_7)^+ = \mathfrak{M} \cup e_7$. Therefore, $e_{fu} = e_5$ is in $(\mathfrak{M} \cup e_7)^+$ and the fault f_u is separable from $f_{y(2)}$.

- Analysis 2: f_u separable from $f_{y(1)}$.

It is necessary to analyze if $e_{fu} \in (\mathfrak{M} \cup e_6)^+$. The Dulmage-Mendelsohn Decomposition of $\mathfrak{M} \cup e_6$ is:

$$\mathcal{DMD}_{\text{IPS(6)}} = \begin{array}{c} \\ e_3 \\ e_4 \\ e_7 \\ e_1 \\ e_2 \\ e_5 \end{array} \begin{array}{ccccc} x_3 & x_4 & x_1 & x_2 & x_5 \\ \left(\begin{array}{ccccc} 1 & 1 & 0 & 0 & 0 \\ 0 & 1 & 1 & 0 & 0 \\ 0 & 0 & 1 & 0 & 0 \\ 0 & 0 & 1 & 1 & 0 \\ 0 & 0 & 1 & 1 & 1 \\ 0 & 0 & 0 & 0 & 1 \end{array} \right) \end{array} \tag{2.65}$$

Matrix $\mathcal{DMD}_{\text{IPS(6)}}$ shows that $(\mathfrak{M} \cup e_6)^+ = \{e_7, e_1, e_2, e_5\}$. Therefore, $e_{fu} = e_5$ is in $(\mathfrak{M} \cup e_6)^+$ and the fault f_u is separable from $f_{y(1)}$.

- Analysis 3: $f_{y(1)}$ separable from $f_{y(2)}$.

It is necessary to analyze if $e_{fy(1)} \in (\mathfrak{M} \cup e_6)^+$. In this case the Dulmage-Mendelsohn Decomposition coincides with $\mathcal{DMD}_{\text{IPS(6)}}$, see Eq. (2.65) in the Analysis 2. Therefore, $e_{fy(1)} = e_7$ is in $(\mathfrak{M} \cup e_6)^+$. As a consequence, the fault $f_{y(1)}$ is separable from $f_{y(2)}$.

- Analysis 4: f_u separable from $f_{y(1)}$ and $f_{y(2)}$.

From applying Definition 2.8 from Sect. 2.3, it is clear that f_u is separable from $f_{y(1)}, f_{y(2)}$ when $e_{fu} \in ((\mathfrak{M} \cup \{e_6, e_7\}) \setminus \{e_{fy(1)}, e_{fy(2)}\})^+$. The biadjacency matrix of $((\mathfrak{M} \cup \{e_6, e_7\}) \setminus \{e_{fy(1)}, e_{fy(2)}\})$ coincides with the biadjacency matrix of $\mathfrak{M}$ because $((\mathfrak{M} \cup \{e_6, e_7\}) \setminus \{e_{fy(1)}, e_{fy(2)}\}) = \mathfrak{M}$. Therefore, its Dulmage-Mendelsohn Decomposition coincides with the presented in matrix $\mathcal{DMD}_{\text{IPS(1)}}$, see Eq. (2.56). Therefore $\mathfrak{M}^+ = [\varPhi]$ and f_u is not separable from $f_{y(1)}$ and $f_{y(2)}$. Therefore, for the IPS under study, in its original description, the separability condition is not reached and the system cannot be diagnosed with FD-IPM.

- Analysis 5: $f_{y(1)}$ separable from f_u and $f_{y(2)}$.
- Analysis 6: $f_{y(2)}$ separable from f_u and $f_{y(1)}$.

Analysis 5 and Analysis 6 are not necessary because it was already shown in Analysis 4 that the separability is not reached for the original model of the IPS. Moreover, it can be shown in a similar way as in Analysis 4 that the separability is not reached.

As a conclusion, for the model of the IPS as described in Sect. 2.4.2, see model in Eq. (2.24), there is no separability among faults with the two current sensors, but the detectability is reached. The IPS does not satisfy hypothesis H3, because it does not satisfy the separability conditions. Therefore, for this benchmark, the faults can only be detected (FD-IPM Steps 1–2).

For the simplified version of the IPS, which corresponds to the Situations 1, 2, and 3, the system can be diagnosed with FD-IPM. For this reason, it is only necessary to make the analysis of the performance of metaheuristics for cases which correspond to these situations.

2.5.3   Benchmark Problem: Two Tanks System

The model that describes the Two Tanks System direct problem, which can be affected by two faults, was presented in Sect. 2.4.3, see Eq. (2.33). The model and its incorporation of the two faults satisfy hypothesis H1. As a first approximation, it is assumed that the leaks in both tanks do not change with time. As a consequence, the system satisfies hypothesis H2.

Considering the optimization problem described by Eq. (2.3) and the restrictions given by Eq. (2.34), it is obtained the following optimization problem, after applying the FD-IPM Step 1:

$$\min \ f\left(\hat{f}_{p(1)}, \hat{f}_{p(2)}\right) = \left\|\sum_{t=1}^{I}\left[L_t\left(f_{p(1)}, f_{p(2)}\right) - \hat{L}_t\left(\hat{f}_{p(1)}, \hat{f}_{p(2)}\right)\right]^2\right\|_{\infty}$$

$$\text{s.t.} \quad \hat{f}_{p(1)}, \hat{f}_{p(2)} \in \mathbb{R} : 0 \le \hat{f}_{p(1)}, \hat{f}_{p(2)} \le 1\,\mathrm{m}^3/s \quad (2.66)$$

where $L_t^T = (L_1(t)\ L_2(t)) \in \mathbb{R}^2$ contains the measurements of the liquid level of each tank at a certain instant of time t; $\hat{L}_t^T = \left(\hat{L}_1(t)\ \hat{L}_2(t)\right) \in \mathbb{R}^2$ contains the calculated values of L at the instant of time t which is obtained from the model of the system, see Eq. (2.33).

The optimization problem in Eq. (2.66) has to be solved in Step 3 of FD-IPM, in order to obtain the fault estimations and provide the result of the diagnosis (FD-IPM Step 4) for this system.

2.5.3.1 Structural Analysis: Verifying Hypothesis H3

Now, it is necessary to verify if the Fault Diagnosis Inverse Problem, formulated as an optimization problem in Eq. (2.66), satisfies hypothesis H3. For that, it is performed the structural analysis of the direct model in Eq. (2.33).

Let's denote by $\mathfrak{M}$ the model of the Two Tanks System. This model has four equations: e_1, e_2, e_3, e_4 and two unknown state variables: L_1, L_2:

$$
\begin{aligned}
&e_1 : \dot{L}_1 = \frac{q_1}{S} - \frac{C_1}{S}\sqrt{L_1} - \frac{C_{12}}{S}\sqrt{|L_1 - L_2|}\,\mathrm{sign}\,(L_1 - L_2) - \frac{f_{p(1)}}{S} \\
&e_2 : \dot{L}_2 = \frac{q_2}{S} - \frac{C_{12}}{S}\sqrt{L_2} + \frac{C_1}{S}\sqrt{|L_1 - L_2|}\,\mathrm{sign}\,(L_1 - L_2) - \frac{f_{p(2)}}{S} \\
&e_3 : y_1 = L_1 \\
&e_4 : y_2 = L_2
\end{aligned}
\tag{2.67}
$$

Each fault only affects one equation:

$$
\begin{aligned}
e_{f_{p(1)}} &= e_1 \\
e_{f_{p(2)}} &= e_2
\end{aligned}
\tag{2.68}
$$

Considering the bipartite graph of the model $\mathfrak{M}$: $G_b = (E; X)$, being $E = \{e_1, \ldots, e_4\}$ and $X = \{L_1, L_2\}$, its biadjacency matrix is:

$$
\mathcal{M}_{2\mathrm{Tanks}(1)} =
\begin{array}{c}
\\ e_1 \\ e_2 \\ e_3 \\ e_4
\end{array}
\begin{array}{cc}
L_1 & L_2 \\
\left(\begin{array}{cc}
1 & 1 \\
1 & 1 \\
1 & 0 \\
0 & 1
\end{array}\right)
\end{array}
\tag{2.69}
$$

From the Dulmage-Mendelsohn Decomposition of the model with biadjacency matrix $\mathcal{M}_{2\mathrm{Tanks}(1)}$, it is obtained that $\mathfrak{M}^+ = \mathfrak{M}$. Therefore, both faults are detectable.

For a separability analysis, it is only necessary to study one case: $f_{p(1)}$ separable from $f_{p(2)}$. Taking into consideration Definition 2.8, the fault $f_{p(1)}$ is separable from $f_{p(2)}$ if:

$$
e_{f_{p(1)}} \in (\mathfrak{M} \backslash e_{f_{p(2)}})^+
\tag{2.70}
$$

in other words, $f_{p(1)}$ is separable from $f_{p(2)}$ if:

$$
e_1 \in (\mathfrak{M} \backslash e_2)^+
\tag{2.71}
$$

The biadjacency matrix for $\mathfrak{M}\backslash e_2$ is:

$$
\mathcal{M}_{2\text{Tanks}(2)} = \begin{array}{c} \\ e_1 \\ e_3 \\ e_4 \end{array} \begin{array}{cc} L_1 & L_2 \\ \begin{pmatrix} 1 & 1 \\ 1 & 0 \\ 0 & 1 \end{pmatrix} \end{array} \tag{2.72}
$$

and its Dulmage-Mendelsohn Decomposition shows that $(\mathfrak{M}\backslash e_2)^+ = M\backslash e_2$. Therefore, $e_1 \in (\mathfrak{M}\backslash e_2)^+$ and $f_{p(1)}$ is separable from $f_{p(2)}$.

As a consequence, the Two Tanks System satisfies the three hypothesis H1, H2, and H3 and it can be diagnosed with FD-IPM.

2.6 Remarks

In this chapter, it was formulated and formalized FDI as a parameter estimation inverse problem. It was also presented an alternative way of determining the uniqueness of the Fault Diagnosis Inverse Problem solutions. All these results are summarized in a methodology for diagnosis: Fault Diagnosis—Inverse Problem Methodology (FD-IPM).

The three chosen benchmark problems that were used during the experiments were also described. The application of FD-IPM Step 1 for each benchmark was also presented.

Chapter 3
Metaheuristics for Optimization Problems

An introduction to metaheuristics for optimization problems is presented in this chapter. In Sect. 3.1 a classification of metaheuristics is put forward. The metaheuristics Differential Evolution (DE); Particle Collision Algorithm (PCA); Ant Colony Optimization (ACO) in its version for continuous problems; and Particle Swarm Optimization (PSO) are described in Sects. 3.2, 3.3, 3.4, and 3.5, respectively. These metaheuristics were applied to the benchmark problems diagnosis described in Chap. 2, based on Fault Diagnosis-Inverse Problem Methodology (FD-IPM) as described in Chap. 2.

Sections 3.6 and 3.7 present two new metaheuristics that are proposed in this book: Particle Swarm Optimization with Memory (PSO-M); and Differential Evolution with Particle Collision (DEwPC).

3.1 Introduction to Metaheuristics for Optimization Problems

An optimization problem consists to find the *best solution* among all possible ones, with respect to some goals. Nowadays, the *best solution* can be interpreted as a solution, which is near enough to the *best solution*, or it is a *good enough* solution, [12].

An optimization problem is continuous, when it is necessary to find the maximum or the minimum of a function with continuous decision variables [12]. Let's denote as f the function at the optimization continuous problem and $\mathbb{D} \subset \mathbb{R}^{n}$ its definition dominion. The description of f is summarized as: $f : \mathbb{D} \subset \mathbb{R}^n \rightarrow \mathbb{R}^m$. If $m > 1$, the optimization problem is multi-objective.

© Springer International Publishing AG, part of Springer Nature 2019
L. Camps Echevarría et al., *Fault Diagnosis Inverse Problems: Solution with Metaheuristics*, Studies in Computational Intelligence 763,
https://doi.org/10.1007/978-3-319-89978-7_3

Let's consider the case when $m = 1$. Due to the fact that:

$$\min f = -\max f \tag{3.1}$$

it is possible to restrict the studies to the problem of finding the minimum of f, without loss of generality.

The function f is called *objective function*.

The optimization problem in Eq. (3.1) may be, or not, subject to equality constraints:

$$g_1(x_1, x_2, \ldots x_n) = 0$$
$$g_2(x_1, x_2, \ldots x_n) = 0$$
$$\vdots \tag{3.2}$$
$$g_p(x_1, x_2, \ldots x_n) = 0$$

or inequality constraints:

$$h_1(x_1, x_2, \ldots x_n) \leq 0$$
$$h_2(x_1, x_2, \ldots x_n) \leq 0$$
$$\vdots \tag{3.3}$$
$$h_r(x_1, x_2, \ldots x_n) \leq 0$$

Metaheuristics for optimization are a group of non-exact algorithms, which allow to solve optimization problems based on a search strategy in the feasible solution space [111, 139]. This search strategy intends to explore the feasible solution space in an efficient-effective way, and it can be easily applied to different optimization problems [111, 112, 138, 139].

Metaheuristics do not provide any guarantee about the quality of the solution. But, they have important advantages, e.g. the ability to look not only for local optima, but also for the global optimum; and the need to evaluate the value of the function only at selected points of the search space without requiring the computation of derivatives [111].

Metaheuristics can be classified considering different criteria and, therefore, there is not a universal classification. With the aim of helping to understand their nature or historical roots, more than their differences, the most common classification features are described in the following subsections.

3.1.1 Classification of Metaheuristics

3.1.1.1 Classification Based on the Number of Solution Candidates at Each Iteration of the Algorithm

- *Population-based Metaheuristics* (PMs):
  These are metaheuristics that at each iteration $Iter$ work with a set of feasible solutions P_{Iter} (Population). The population from the previous iteration, P_{Iter-1}, is used for generating P_{Iter}. The scheme of these methods can be summarized as:

$$P_{Iter} = h(F(P_{Iter-1})) \tag{3.4}$$

where F is a *fitness* function that evaluates the quality of each member of the population; and h is a function that handles the current population in order to generate a new one [3].

Examples of PMs are:

- Evolutionary Algorithms (EAs), [4].
- Swarm Intelligence Algorithms (SIs), [14].
- Algorithms based on physical processes such as Multiple Particle Collision Algorithm (MPCA), [87].
- Memetic Algorithms (MAs), [52, 145].

- *Non-Population-based Metaheuristics* (NPMs):
  These are metaheuristics that at each iteration $Iter$ work with only one feasible solution X_{Iter-1} for generating a new one X_{Iter}. The scheme of these methods can be summarized as:

$$X_{Iter} = h(X_{Iter-1}) \tag{3.5}$$

where h is a function that handles the current feasible solution X_{Iter-1} in order to generate a new one [3].

Examples of NPMs are:

- Simulated Annealing (SA), [73].
- Tabu Search (TS), [46].
- Particle Collision Algorithm (PCA), [104].

Let's note that the sets PMs and NPMs are mutually exclusive, it means that $PMs \bigcap NPMs = \Phi$

Classification of Population-based Metaheuristics (PMs):

- *Evolutionary Algorithms (EAs):*
  These algorithms simulate the principles of the Theory of Evolution: election, mutation, reproduction, and selection. These principles are incorporated in the function h, which is described as:

$$h(F(P_{Iter-1})) = \mu(s(P_{Iter-1})) \tag{3.6}$$

where μ is called *mutation* function; and s is the *selection* function [3].

At the beginning, these methods were all inspired by the well-known evolutionary mechanisms from nature, being, therefore, also known as bio-inspired algorithms (e.g., Genetic Algorithms were inspired by the Theory of Evolution of the Species proposed by Darwin). With the hybridization of algorithms, new evolutionary mechanisms, which are not bio-inspired were introduced. As a consequence, EAs is divided into two big groups: Bio-inspired and non-bio-inspired.

Examples of EAs are:

- Genetic Algorithms (GAs), [47].
- Differential Evolution (DE), [100].
- Evolutionary Programming (EP) [3, 4].
- Genetic Programming (GP), [47].
- Biogeography-Based Optimization (BBO), [49, 50].
- Generalized Extremal Optimization (GEO), [121].
- Estimation of Distribution Algorithms (EDAs), [79].

- *Memetics Algorithms (MAs):*
  Memetics Algorithms were developed in the 80s of the twentieth century. They are similar to EAs, but in this case, the algorithms are inspired by the Theory of the Cultural Evolution [52, 145] and the *gens* from EAs are substituted by *memes*. The concept of meme was introduced by R. Dawkins in the 70s [28]: *A meme represents a unit of cultural evolution that can exhibit local refinement.*

  In MAs, the members of the population *compete* and *cooperate* among them [52, 145], in order to improve the quality of the population. Here the concept of *imitation* plays an important role for generating the population. As a consequence, function h in Eq. (3.6) incorporates these principles.

  MAs can also be understood as extensions of EAs, in which are applied separate processes in order to refine individuals [75].

- *Swarm Intelligence Algorithms (SIs):*
  The term Swarm Intelligence (SI) was introduced by Beni and Wang [11], in the late 80s of the past century in the context of cellular robotic systems. However, SI has been extended and nowadays it involves algorithms that model the collective behavior of social agents in nature, such as ant colonies, honey bees, and bird flocks [14]. Therefore, these algorithms are Population-based algorithms [3].

  In SIs, the function h is based on a model that describes certain auto organization characteristics within groups of agents [3]. For that reason, many authors do not consider that Swarm Intelligence Algorithms are Evolutionary Algorithms (the algorithms in SI do not generate a populations based on the basic operators from EAs: election, mutation, reproduction, and selection).

  On the other hand, other authors consider that Swarm Intelligence Algorithms are evolutionary. In Ref. [37], it is presented a comparison between GAs and the scheme of Particle Swarm Optimization (PSO, one of the classic algorithms from SI), with the aim of showing the *evolutionary* characteristics in PSO.

Examples of SIs are

- Ant Colony Optimization (ACO), [13, 33, 34, 119].
- Particle Swarm Optimization (PSO), [69, 71].
- Artificial Bee Colony (ABC), [67].
- Cuckoo Search (CS), [139–141].
- Firefly Algorithm (FA), [138, 139].
- Intelligent Water Drops algorithm (IWD), [108].

3.1.1.2 Classification Based on the Metaheuristic Inspiration Source

The metaheuristics may be inspired or not by processes or phenomena observed in nature. Therefore, they may be classified based on their inspiration source into two groups: Nature-inspired Metaheuristics and those which are not Nature-inspired. When the processes or phenomena from the nature are biological, it is said that the metaheuristics are Bio-inspired.

This classification is summarized as follows:

- *Non-Nature-inspired Metaheuristics (NNMs):* They may be derived simply from mathematical operations, with no inspiration in nature mechanisms.

  Examples of NNMs are:

  - Tabu Search (TS), [46].
  - Differential Evolution (DE), [100].

- *Nature-inspired Metaheuristics (NMs):* They are based on processes or phenomena from nature.

Examples of NMs are:

- Bio-inspired Metaheuristics (BMs).
- Metaheuristics inspired on physical phenomena: SA, PCA.
- Metaheuristics inspired on cultural phenomena: MAs.

- *Bio-inspired Metaheuristics (BMs):* Metaheuristics based on biological mechanisms.

Examples of BMs are:

- Most of EAs: GAs, GEO, BBO, and GP.
- SI algorithms: ACO, PSO, ABC, CS, and FA.

The relation between BMs and NMs is expressed as:

$$BMs \subset NMs \tag{3.7}$$

In Fig. 3.1 a diagram with the metaheuristics previously mentioned and their classification based on the described criteria is presented

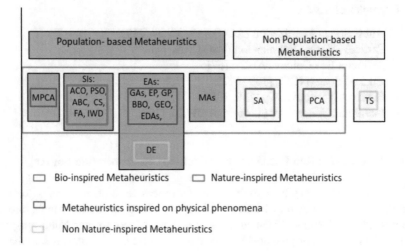

Fig. 3.1 Proposed classification of the metaheuristics mentioned in this chapter

3.2 Differential Evolution (DE)

The methods denominated Differential Evolution (DE) is an evolutionary algorithm proposed in 1995 for optimization problems. Since the beginning, it was described by its authors as *a simple and efficient adaptive scheme for global optimization over continuous spaces* [123]. Some of its most important advantages are: simple structure, simple computational implementation, speed and robustness [91, 100, 123, 124].

The metaheuristic DE is based on three operators: *Mutation*, *Crossover*, and *Selection* [123, 124]. These operators have the same names as the operators from GA, but they are based on vector operations [100, 111, 124]. For this reason, many authors do not consider it as a bio-inspired algorithm.

3.2.1 Description of the DE Algorithm

The algorithm generates at each iteration $Iter$ a new population of Z feasible candidate solutions $P_{Iter} = \left\{ X_{Iter}^1, X_{Iter}^2, \ldots, X_{Iter}^Z \right\}$ with the application of the three operators to the previous population P_{Iter-1}.

The algorithm needs to be initialized, so the initial population $P_0 = \left\{ X_0^1, X_0^2, \ldots, X_0^Z \right\}$ is formed by Z feasible candidate solutions for the problem by Eqs. (3.1–3.4). The candidate solutions are vectors in $\mathbb{R}^n$.

At every iteration, the operators *Mutation*, *Crossover*, and *Selection* are applied. First Mutation is applied over the population of the previous iteration, and a candidate population of Z vectors is generated: $\hat{P}_{Iter} = \left\{ \hat{X}^1_{Iter}, \hat{X}^2_{Iter}, \ldots, \hat{X}^Z_{Iter} \right\}$. To the population obtained from Mutation, it is applied the Crossover. The Crossover operator incorporates components with a certain probability, from a candidate vector $\hat{X}^z_{Iter}$ to the corresponding vector X^z_{Iter-1}. After the Crossover, it is obtained a population of Z vectors, which are usually called *trial vectors*. Finally, the operator *Selection* is applied in order to decide which of the Z trial vectors obtained after the Crossover will be part of the population generated at the current iteration. With the new population $P_{Iter} = \left\{ X^1_{Iter}, X^2_{Iter}, \ldots, X^Z_{Iter} \right\}$, the best solution is updated. This scheme is executed until the best solution provided by the algorithm is considered *good enough* according to a prescribed stopping criterion.

Following, the most used *Mutation* operators are described:

- $X^\delta_{Iter-1}/1$: one pair of solutions is used for perturbing another solution X^δ_{Iter-1} from population P_{Iter-1}. The solutions are randomly selected. This mechanism is executed Z times, in order to obtain Z candidate solutions after Mutation: $\hat{X}^z_{Iter}$, with $z = 1, \ldots, Z$,

$$\hat{X}^z_{Iter} = X^\delta_{Iter-1} + C_{\text{scal}} \left(X^\alpha_{Iter-1} - X^\beta_{Iter-1} \right) \tag{3.8}$$

being $X^\delta_{Iter-1}, X^\alpha_{Iter-1}, X^\beta_{Iter-1} \in \mathbb{R}^n$ solutions from P_{Iter-1}; and C_{scal} a parameter called *Scaling factor*.

- $X^{\text{best}}/1$: one pair of solutions is used for perturbing the current best solution X^{best}. This pair of solutions is randomly selected. This mechanism is executed Z times, in order to obtain Z candidate solutions after Mutation: $\hat{X}^z_{Iter}$, with $z = 1, \ldots, Z$,

$$\hat{X}^z_{Iter} = X^{\text{best}} + C_{\text{scal}} \left(X^\alpha_{Iter-1} - X^\beta_{Iter-1} \right) \tag{3.9}$$

where $X^{\text{best}}, X^\alpha_{Iter-1}, X^\beta_{Iter-1} \in \mathbb{R}^n$ are solutions from P_{Iter-1}.

- $X^{\text{best}}/2$: two pairs of solutions are used for perturbing the current best solution X^{best}. These two pairs of solutions are randomly selected. This mechanism is executed Z times, in order to obtain Z candidate solutions after Mutation: $\hat{X}^z_{Iter}$, with $z = 1, \ldots, Z$,

$$\hat{X}^z_{Iter} = X^{\text{best}} + C_{\text{scal}} \left(X^{\alpha 1}_{Iter-1} - X^{\alpha 3}_{Iter-1} + X^{\alpha 2}_{Iter-1} - X^{\alpha 4}_{Iter-1} \right) \tag{3.10}$$

where $X^{\text{best}}, X^{\alpha 1}_{Iter-1}, X^{\alpha 2}_{Iter-1}, X^{\alpha 3}_{Iter-1}, X^{\alpha 4}_{Iter-1} \in \mathbb{R}^n$ are solutions from P_{Iter-1}.

After the Mutation, the **Crossover** operator is applied according to the following description:

Each component $\hat{x}^z_{(Iter)i}$ of the candidate solution $\hat{X}^z_{Iter}$ with $i = 1, 2, .., n$ and $z = 1, 2, \ldots, Z$, which has been previously generated by *Mutation*, goes through the following procedure:

$$\hat{x}^z_{(Iter)i} = \begin{cases} \hat{x}^z_{(Iter)i} & \text{if } q_{rand} \leq C_{\text{cross}} \\ x^\delta_{(Iter-1)i} & \text{otherwise} \end{cases} \tag{3.11}$$

where $0 \leq C_{\text{cross}} \leq 1$ is the *Crossover Factor*, another parameter from DE; q_{rand} is a random number, which is generated by means of the distribution function represented by λ in the generation mechanism.
The Z solutions obtained after the Crossover are called *trial vectors*.

The **Selection** operator has the following description:
For the vector X^z_{Iter}, to be part of the new population, it is selected following the rule:

$$X^z_{Iter} = \begin{cases} \hat{X}^z_{Iter} & \text{if } f(\hat{X}^z_{Iter}) \leq f(X^\delta_{Iter-1}) \\ X^\delta_{Iter-1} & \text{otherwise} \end{cases} \tag{3.12}$$

The difference within the generation mechanisms lead commonly in differences in the operator *Mutation*. The most used generation mechanisms, which correspond to the three schemes for Mutation described in this section, are:

- DE/X^δ_{Iter-1}/1/*bin*: where *bin* is the notation for a binomial distribution function, that is used during the Crossover.
- DE/X^{best}/1/*bin*
- DE/X^{best}/2/*bin*

The control parameters in DE are: size of the population, Z; Crossover Factor, C_{cross}; and Scaling Factor, C_{scal}. In the algorithms presented in this chapter, *MaxIter* represents the maximum number of iterations to be performed.

Some modifications on the original version of DE have been made, in order to improve its ability on escaping from local minima. The more successful variants of DE are focused on changes performed on the *Mutation* operator and in the self adaptation of parameters C_{cross} and C_{scal} [15, 27, 91, 102, 129, 144, 146, 148]. The method DE has also been hybridized with other metaheuristics [50].

A general description of the algorithm for DE is given in Algorithm 1. This description is based on the original version, which was proposed in Refs. [123, 124].

3.2.2 Remarks on DE

- The search directions and the possible step sizes depend on the candidate solutions used in *Mutation*.
- The parameter C_{cross} determines the influence of X^δ_{Iter-1}: a higher value of this parameter indicates a lower influence.

Algorithm 1: Algorithm for the Differential Evolution (DE)

Data: C_{cross}, C_{scal}, Z, $MaxIter$, Generation Mechanism
Result: X^{best}
Generate an initial random population of Z solutions;
Select the best solution X^{best};
for $Iter = 1$ *to* $Iter = MaxIter$ **do**
 | Execute *Mutation*, Eq. (3.10);
 | Execute *Crossover*, Eq. (3.11);
 | Execute *Selection*, Eq. (3.12);
 | Update X^{best};
 | Verify stopping criteria;
end
Solution: X^{best}

- The parameter C_{scal} determines the influence of the vectors used in the *Mutation* [91].
- Simple rules for selecting values for C_{cross} and C_{scal} depending on the application considered are given in [124].
- A better diversification in the search can be obtained by means of increasing the number of pair of vectors used by the *Mutation* operator and also the size of the population, Z. However, this may cause a decrease in the probability of finding the right search direction [91].

3.3 Particle Collision Algorithm (PCA)

The metaheuristic Particle Collision Algorithm (PCA) was developed with the aim to improve some of the disadvantages of the metaheuristic entitled Simulated Annealing (SA). It is based on the particle collision phenomena inside a nuclear reactor [104, 111].

Initially, it was developed as a non-population-based metaheuristic, but after it was extended to populations. This new variant is called *Multiple Particle Collision Algorithm* (MPCA), [87].

PCA is also a member of the well-known class of *Metropolis Algorithms* [90]. It means that during its random search, it can accept, at any iteration, a solution that does not improve the objective function [111].

3.3.1 Description of the PCA Algorithm

In this section, the original version of PCA is presented [104, 111]. PCA starts with an initial solution X_0 and it generates a new feasible candidate solution $\hat{X}_{Iter}$ at every iteration $Iter$. The generation scheme is based on stochastic modifications on the previous candidate solution X_{Iter-1}.

Inside a nuclear reactor two situations may occur after the interaction between a particle, the neutron, and a nucleus: the particle may be absorbed or scattered. These interactions are computationally simulated, and this is used as the basis for the decision of acceptance or not of the new feasible candidate solution $\hat{X}_{Iter}$ as the new solution X_{Iter} [104, 111]: the absorption means that the new feasible candidate solution $\hat{X}_{Iter}$ is accepted as the new solution X_{Iter}; and the scattering is identified with its rejection, followed by the random generation of a new candidate solution.

The *Absorption-Scattering* operator is explained next:

- *Absorption*:

  If $f(\hat{X}_{Iter}) \leq f(X_{Iter-1})$, then $\hat{X}_{Iter}$ is absorbed (accepted as new feasible solution X_{Iter}); and a random locally restricted search by means of an operator called *Local* is made around it. It can be summarized as follows:

$$f(\hat{X}_{Iter}) \leq f(X_{Iter-1}) \Longrightarrow X_{Iter} = \hat{X}_{Iter} \qquad (3.13)$$

  and operate with $Local(X_{Iter})$

- *Scattering with probability $p_{r(Iter)}$*:

  In order to escape from eventual local optima, the rejection of a candidate solution is designed as a rejection with a probability $p_{r(Iter)} = 1 - \dfrac{f(X^{\text{best}})}{f(\hat{X}_{Iter})}$.

  This means that even when the new feasible solution $\hat{X}_{Iter}$ does not improve the value of the objective function, it may be accepted with a certain probability $p_{r(Iter)}$. This probability depends on the values of the objective function for the best current solution and of the new candidate solution for the current iteration $Iter$.

  If $f(\hat{X}_{Iter}) > f(X_{Iter-1})$, then the new feasible solution $\hat{X}_{Iter}$ is not immediately rejected. Instead, PCA generates a random number q and two possible situations may occur:

  - If $q < p_{r(Iter)}$, then *Absorption* is applied to $\hat{X}_{Iter}$.
  - If $q \geq p_{r(Iter)}$, then $\hat{X}_{Iter}$ is rejected, and a new candidate solution is randomly generated.

The algorithm of *Absorption-Scattering* is presented in Algorithm 2. In this algorithm $MaxIter_c$ stands for the maximum number of iterations to be performed for the operator *Local*.

The operator *Local* makes $MaxIter_c$ consecutive stochastic perturbations of the same order, to every component of a certain feasible solution X. Being X_{Iter_c} the vector, that results from applying the operator *Local* to X in iteration $Iter_c$ of this operator; denoting by x_i, with $i = 1, 2, \ldots, n$, the components of the vector X; and $x_{(Iter_c)i}$, with $i = 1, 2, \ldots, n$, the components of X_{Iter_c}, the perturbations are described as:

$$x_{(Iter_c)i} = x_{(Iter_c-1)i} + (x^U_{(Iter_c)i} - x_{(Iter_c-1)i})q_1$$
$$-(x^L_{(Iter_c)i} - x_{(Iter_c-1)i})(1 - q_1) \qquad (3.14)$$

Algorithm 2: Algorithm for the *Absorption-Scattering* operator of the Particle Collision Algorithm (PCA)

Data: $\hat{X}_{Iter}$, X^{best}, X_{Iter-1}, $MaxIter_c$
Result: Solution: X_{Iter}
if $f(\hat{X}_{Iter}) \leq f(X_{Iter-1})$ **then**
   | $X_{Iter} = Local(\hat{X}_{Iter}, MaxIter_c)$;
else
   | Generate a random number q;
   | **if** $q < 1 - \frac{f(X^{best})}{f(\hat{X}_{Iter})}$ **then**
   | | $X_{Iter}=Local(\hat{X}_{Iter}, MaxIter_c)$;
   | **else**
   | | Randomly generate a new candidate solution X_{Iter};
   | **end**
end

---

being q_1 a random number in $[0; 1]$, and:

$$x^L_{(Iter_c)i} = \begin{cases} q_2 \, x_{(Iter_c-1)i} & \text{if } q_2 \, x_{(Iter_c-1)n} \geq L_i \\ L_i & \text{otherwise} \end{cases} \tag{3.15}$$

$$x^U_{(Iter_c)i} = \begin{cases} q_3 \, x_{(Iter_c-1)i} & \text{if } q_3 \, x_{(Iter_c-1)i} \leq U_i \\ U_i & \text{otherwise} \end{cases} \tag{3.16}$$

where L_i and U_i are the minimum and the maximum values in the search space, respectively, for the variable, that corresponds to the component x_i; q_2 is a random number in the range $[0.8; 1.0]$; and q_3 is also a random number in the range $[1.0; 1.2]$, [104, 105].

The algorithm for the *Local* operator is represented in Algorithm 3.

The metaheuristic PCA has only two control parameters: the maximum number of iterations *MaxIter* of the main loop of the algorithm; and the maximum number of iterations $MaxIter_c$ of *Absorption-Scattering* operator, as already shown in Algorithms 2 and 3. The original version of PCA is represented in Algorithm 4 [104].

3.3.2 Remarks on PCA

- The search space is explored based on consecutive absorption-scattering events [104, 111].
- The operator *Local* allows to make an intensification of the search, i.e. exploitation.
- The algorithm has only two control parameters. That makes it feasible, and robust, for a great number of applications.

Algorithm 3: Algorithm for the *Local* operator of the Particle Collision Algorithm (PCA)

Data: X, $MaxIter_c$
Result: Solution after search: X
$X^0 = X$;
for $l = 1$ *to* $l = MaxIter_c$ **do**
　　Random generation of $X^{l'}$ based on Eq. (3.14);
　　if $f(X^{l'}) \leq f(X^{l-1})$ **then**
　　　| $X^l = X^{l'}$;
　　end
　　$X^l = X^{l-1}$;
end
$X = X^{MaxIter_c}$;

Algorithm 4: Algorithm for the Particle Collision Algorithm (PCA)

Data: $MaxIter$, $MaxIter_c$
Result: Best solution: X^{best}
Generate a random initial solution X_0;
$X^{best} = X_0$;
for $l = 1$ *to* $l = MaxIter$ **do**
　　Generate $\hat{X}_l$ within the search domain;
　　Apply *Absorption-Scattering*$(\hat{X}_l, X^{best}, X_{l-1}, MaxIter_c)$;
　　Update X^{best};
　　Verify stopping criteria;
end
Best Solution: X^{best}

- The values of the objective function f used in different steps of PCA may be substituted for values of a certain fitness function F, according to the problem of interest to be solved.

3.4   Ant Colony Optimization (ACO)

The metaheuristic entitled Ant Colony Optimization (ACO) was proposed by Marco Dorigo in 1992, as part of the results of his doctoral thesis [32]. It was initially proposed for integer programming problems [33, 34], but it has been successfully extended to continuous optimization problems [13, 111, 118, 119].

The metaheuristic ACO is inspired on the behavior of ants, specifically when they are seeking a path between their colony and a food source. It is observed in nature that after a certain time the ants of a colony are able to find the shortest path between these two positions [13, 33, 34, 113, 119]. This behavior is explained with the deposition and evaporation of a substance: the pheromone [13, 33, 34, 113, 119]. The ways with greater concentration of this substance attract the other ants; and they will deposit their pheromone in these ways.

The idea of the deposit and evaporation of the pheromone is simulated by means of a pheromone matrix $\mathbb{F}$, which is updated at each iteration and is accessible to all the ants in the colony [13, 33, 34, 119].

In ACO, each ant is associated to a feasible candidate solution and the best solution is the *best ant*.

3.4.1 Description of the ACO Algorithm for Continuous Problems

This subsection describes the adaptation of ACO to the continuous case that was reported in [111, 122]. This variant was successfully applied to other problems [7, 122].

At each iteration $Iter$, ACO generates a new colony with Z ants: $P_{Iter} = \{X^1_{Iter}, X^2_{Iter}, \ldots, X^Z_{Iter}\}$. Each ant X^z_{Iter}, with $z = 1, 2, \ldots, Z$ represents a feasible solution. Therefore, each ant is a vector of n components $X^z_{Iter} = [x^{(z)}_{(Iter)1} \ x^{(z)}_{(Iter)2} \ \ldots x^{(z)}_{(Iter)n}]^T$. The generation of P_{Iter} is based on a probability matrix PC which depends on $\mathbb{F}$.

In this variant, the first step consists of discretizing the feasible interval of each decision variable x_i, with $i = 1, 2, \ldots, n$, of the problem into K possible values x^k_i, with $k = 1, 2, \ldots, K$. The generation of the ants paths at each iteration uses the information that was obtained from the previous paths. The information is saved in the pheromone accumulative probability matrix $PC \in \mathbb{M}_{n \times K}(\mathbb{R})$, whose elements are updated at each iteration $Iter$ as a function of $\mathbb{F} \in \mathbb{M}_{n \times K}(\mathbb{R})$:

$$pc_{ij}(Iter) = \frac{\sum_{l=1}^{j} f_{il}(Iter)}{\sum_{l=1}^{K} f_{il}(Iter)} \tag{3.17}$$

being f_{ij} elements of $\mathbb{F}$ and they express the pheromone level of the jth discrete value of the ith decision variable.

The elements of $\mathbb{F}$ are also updated at each iteration based on the evaporation factor C_{evap}, and on the incremental factor C_{inc}:

$$f_{ij}(Iter) = (1 - C_{evap})f_{ij}(Iter - 1) + \delta_{ij,\text{best}}C_{inc}f_{ij}(Iter - 1) \tag{3.18}$$

being

$$\delta_{ij,\text{best}} = \begin{cases} 1 \text{ if} & x^j_i = x^{(\text{best})}_i \\ 0 \text{ otherwise} \end{cases} \tag{3.19}$$

and $x^{(\text{best})}_i$ is the ith component of the best solution X^{best}.

Matrix $\mathbb{F}$ is randomly initialized. All its elements take random values within the interval $[0; 1]$.

Algorithm 5: Algorithm for the Ant Colony Optimization (ACO)

Data: C_{evap}, C_{inc}, q_0, K, Z, $MaxIter$
Result: Best solution: X^{best}
Discretize the interval for each variable in K values;
Generate a random initial pheromone matrix $\mathbb{F}$ with the same value for all the elements f_{ij};
Compute matrix PC based on Eq. (3.17);
Generate the initial set of ants paths with Eq. (3.20) and update X^{best};
for $l = 1$ *to* $l = MaxIter$ **do**
$\quad$ Update matrix $\mathbb{F}$ based on Eq. (3.18);
$\quad$ Update matrix PC based on Eq. (3.17);
$\quad$ Generate a new set of ants paths with Eq. (3.20);
$\quad$ Update X^{best};
$\quad$ Verify stopping criteria;
end
Solution: X^{best}

The scheme for generating the new ants paths considers a prescribed parameter q_0. Each zth ant to be generated uses the following scheme:

- Generate n random numbers $q_1^{rand}, q_2^{rand}, \dots, q_n^{rand}$.
- The value of the ith component, with $i = 1, 2, \dots, n$, of the zth ant is:

$$x_i^{(z)} = \begin{cases} x_i^{\bar{m}} & \text{if } q_i^{rand} < q_0 \\ x_i^{\hat{m}} & \text{if } q_i^{rand} \geq q_0 \end{cases} \qquad (3.20)$$

where $\bar{m} : \quad f_{i\bar{m}} \geq f_{im} \ \forall \ m = 1, 2, \dots, K$ and $\hat{m} : \quad \left(pc_{i\hat{m}} > q_i^{rand}\right) \wedge (pc_{i\hat{m}} \leq pc_{im}) \ \forall m \geq \hat{m}$

The control parameter q_0 allows controlling the level of randomness during the ant generation. This fact determines, jointly to Z and K, the level of search intensification or diversification in ACO [113]. At the end of each iteration, the best solution X^{best} is updated.

The pseudo-code for this version of ACO is given in Algorithm 5.

3.4.2 Ant Colony Optimization with Dispersion (ACO-d)

Ant Colony Optimization with Dispersion, ACO-d, intends to simulate a pheromone deposit, which is nearer to the real: the pheromone affects the path where it is deposited, and the paths next to it [8]. This is called dispersion. Another variants of ACO are also based on changes to the pheromone updating [107].

The dispersion is established as a variation in the updating equation (Eq. (3.18)), of the elements in matrix $\mathbb{F}$. As a consequence, a new parameter is introduced: coefficient of dispersion, C_{dis}.

Therefore, the difference between ACO and ACO-d is based on the way the pheromone matrix is updated [8]. In [8] this scheme is applied to the traveling salesman problem. In the present work, ACO-d is adapted to continuous problems.

3.4.2.1 Adaptation of ACO-d to Continuous Problems

The pheromone deposition considers the scheme described by Eq. (3.18), and it also includes an extra deposit (dispersion) of the pheromone in solutions that are close to the best one X^{best}. Therefore, the first step is to define the maximum number of neighbors of X^{best}, which will receive an extra deposit of pheromone. For that, it is adopted a scheme for which each component x_i^{best}, with $i = 1, 2, \ldots, n$, has a maximum number of neighbors for receiving pheromone. Let's denote such set of neighbors $V[x_i^{\text{best}}]$ and let's introduce its definition:

Definition 3.1 Let's denote by X^{best} the vector that represents the best solution provided by ACO-d until a given iteration, and by x_i^{best} the ith component of this solution. The Neighborhood for dispersion of x_i^{best}, denoted by $V[x_i^{\text{best}}]$, is defined as:

$$V[x_i^{\text{best}}] = \left\{ x_i^m : d\left(x_i^{\text{best}}, x_i^m\right) < d_{\max}, \ 1 < m \le K \right\} \tag{3.21}$$

The distance $d_{\max}$, that appears in Definition 3.1, is computed taking the average of the half of all the possible distances between values x_i^m, and x_i^r with $m, r = 1, 2, \ldots K$, in ascending order [8]. This adaptation of ACO-d to the continuous optimization case is based on the version of ACO that was described in the previous Sect. 3.4.1. As a consequence of the discretization, the value for $d_{\max}$ is computed by means of:

$$d_{\max} = \frac{h + 2h + 3h + \ldots + \left[\frac{K}{2}\right]h}{\left[\frac{K}{2}\right]} \tag{3.22}$$

where $h = \frac{b-a}{K}$ with $x_i \in (a, b)$, and $[x]$ represents the nearest integer to x.

Working with Eq. (3.22), and considering the sum of the n first integers, this equation can be reformulated as:

$$d_{\max} = h \frac{\left[\frac{K}{2}\right] + 1}{2} \tag{3.23}$$

Noticing that $d\left(x_i^m, x_i^{m+1}\right) = h$, then Definition 3.1 is expressed as:

$$V[x_i^{\text{best}}] = \left\{ x_i^m : d\left(x_i^{\text{best}}, x_i^m\right) < \frac{\left[\frac{K}{2}\right] + 1}{2}, \ 1 < m \le K \right\} \tag{3.24}$$

Making $x_i^m = a+hm$ and $x_i^{best} = a+h\bar{m}$, Definition 3.1 can also be expressed as:

$$V[x_i^{best}] = \left\{ x_i^m : \ \bar{m} - \frac{\left|\frac{K}{2}\right| + 1}{2} < m < \bar{m} + \frac{\left|\frac{K}{2}\right| + 1}{2}, \ 1 < m \leq K \right\} \qquad (3.25)$$

Therefore, the scheme for the pheromone deposit for $x_i^m \in V[x_i^{best}]$ is expressed as:

$$f_{im}(Iter) = f_{im}(Iter) + \frac{C_{dis}}{\bar{m} - m} \qquad (3.26)$$

where $f_{im}(Iter)$ is the value from ACO, see Eq. (3.18).

3.4.3 Hybrid Strategy Ant Colony Optimization with Differential Evolution (ACO-DE)

This hybridization was proposed after analyzing the results of the experiments with the Two Tanks system, see Chap. 4. During the experiments, it was observed that the quality of the estimations of DE was very dependent on its initial population. It was also observed that ACO was able to give reasonable good estimations in a few number of iterations. As a consequence, the idea of hybridizing both algorithms in order to reduce the number of objective function evaluations for DE, together with improving the accuracy of estimations for DE emerged [19]. The idea of the hybridization may be summarized as using ACO for obtaining a good initial population for DE. The hybrid ACO-DE was then originally proposed for continuous problems [19] and its description is as follows:

Apply ACO with a small K value, i.e. using a not refined mesh for the discrete values for each variable, reducing then the possible values of the variables, and therefore, the search space. Subsequently, it is used the best ants of the history of ACO (the number of ants could be a fraction of the total number of ants) as the initial population of DE, with the aim of performing an intensification around an already identified promising area. In summary:

1. Apply ACO with a small value of K, see algorithm in Algorithm 5, in order to make a first rapid search.
2. Use the better ants of the colony obtained at the end of the application of ACO (from previous step) as the initial population for DE. Apply DE, see algorithm in Algorithm 1.

The validation process for this approach is not presented because it is just a hybridization, not a new metaheuristic. This algorithm is just the application of ACO and DE, without changes on their structure, in a sequential procedure.

3.4.4 Remarks on ACO

- An advantage of ACO is that its parameters may be manipulated in order to achieve more diversification or intensification during the search. This allows an efficient hybridization with other algorithms.
- The hybrid strategy ACO-DE is inspired by the study of the evolution of the best objective function value for DE and ACO, see Sect. 4.4.

3.5 Particle Swarm Optimization (PSO)

Particle Swarm Optimization is an algorithm classified within the field of Swarm Intelligence. It was introduced by Kennedy and Eberhart in 1995 [69, 71] for solving optimization problems, and it is based on the social behavior of birds flocks and fishes schools [69, 71, 113]. It is a population-based algorithm [14].

PSO has been applied to different fields requiring optimization in high dimensional spaces. This is a result of its simplicity, high efficiency in searching, easy implementation, and its fast convergence to the global optimum [65].

3.5.1 Description of the PSO Algorithm

PSO considers a group or population (swarm) of Z agents or particles (birds), which are dedicated to finding a good approximation to the global minimum in a search space. Each particle moves throughout the search space $\mathbb{D}$. The position of the zth particle, in iteration $Iter$, is identified with a solution $X^z_{Iter} \in \mathbb{R}^n$ of the optimization problem. At each iteration, all the particles update their positions.

In order to update its position, each particle *saves* its best historical position $X^{z(pbest)}$. This is a way to accumulate its individual experience. Each particle also has access to the collective experience, which is represented by the best historical position for the whole swarm: X^{gbest}.

The scheme for generating the new position for each particle is:

$$X^z_{Iter} = X^z_{Iter-1} + V^z_{Iter} \tag{3.27}$$

being X^z_{Iter-1} the position of this particle in the previous iteration $Iter - 1$; and $V^z_{Iter} \in \mathbb{R}^n$ its velocity at the current iteration.

The vector V^z_{Iter} is also updated at each iteration:

$$V^z_{Iter} = V^z_{Iter-1} + c_1 \Xi \left(X^{z(pbest)} - X^z_{Iter-1} \right)$$
$$+ c_2 \Xi \left(X^{gbest} - X^z_{Iter-1} \right) \tag{3.28}$$

being V_{Iter-1}^z the velocity of this particle in the previous iteration; Ξ denotes a diagonal matrix with random numbers in the interval $[0;1]$; and c_1, c_2 are parameters, which characterize the trend during the velocity updating [65, 69].

Parameters c_1 and c_2 (cognitive and social parameter, respectively) represent how the individual and social experience influence in the next particle decision. Some studies have been made in order to determine the best values for c_1 and c_2. The values $c_1 = c_2 = 2$, $c_1 = c_2 = 2.05$ or $c_1 > c_2$ with $c_1 + c_2 \leq 4.10$ are recommended [10, 23, 70].

The main difference within the variants of PSO is in updating the position and velocity of each particle. Equations (3.27) and (3.29) represent the canonical algorithm. A well-known variant is the one with inertial weight. The idea behind this variant is to add an inertial factor ω for balancing diversification and intensification of the search [10, 65]. The inertial weight can be constant or not. This parameter ω affects the updating velocity through the expression:

$$V_{Iter}^z = \omega V_{Iter-1}^z + c_1 \Xi \left(X^{z(pbest)} - X_{Iter-1}^z \right) \tag{3.29}$$
$$+ c_2 \Xi \left(X^{gbest} - X_{Iter-1}^z \right)$$

It is usual to start with $\omega = 1$, and to reduce its value by an exponential law [65]. Nowadays, the most accepted strategy for ω is to establish $\omega \in [\omega_{min}; \omega_{max}]$ and to reduce its value according to the current iteration number $Iter$ by means of:

$$\omega = \omega_{max} - \frac{\omega_{max} - \omega_{min}}{MaxIter} Iter \tag{3.30}$$

where $MaxIter$ is the maximum number of iterations to be reached. It is recommended to take $\omega_{max} = 0.9$ and $\omega_{min} = 0.4$ [81].

In order to control the velocity, sometimes it is added a new parameter called constriction factor χ. The new expression for updating the velocity [26, 65] then written as:

$$V_{Iter}^z = \chi \left[V_{Iter-1}^z + c_1 \Xi \left(X^{z(pbest)} - X_{Iter-1}^z \right) \right. \tag{3.31}$$
$$\left. + c_2 \Xi \left(X^{gbest} - X_{Iter-1}^z \right) \right] \tag{3.32}$$

being

$$\chi = \frac{2}{\left| 2 - \varphi - \sqrt{\varphi^2 - 4\varphi} \right|} \tag{3.33}$$

and $\varphi = c_1 + c_2 > 4$.

Algorithm 6: Algorithm of the Particle Swarm Optimization (PSO)

Data: c_1, c_2, Z, $MaxIter$
Result: Best solution: X^{best}
Compute χ with Eq. (3.33);
Generate randomly X_0^z and V_0^z for Z particles;
Update $X^{z(pbest)}$ and X^{gbest};
for $l = 1$ to $l = MaxIter$ **do**
  for $z = 1$ to $z = Z$ **do**
    Update V_l^z with Eq. (3.32);
    Update X_l^z with Eq. (3.27);
    Update $X^{z(pbest)}$;
  end
  Update X^{gbest};
  Verify stopping criteria;
end
Solution: X^{gbest}

The values $c_1 = c_2 = 2.05$, and therefore $\chi = 0.729$, are recommended in the literature [38]. Moreover, from Eqs. (3.30) and (3.32), it is obtained that these values are equivalent to use constant inertial weight with $\omega = 0.729$, and $c_1 = c_2 = 1.49$ [38].

There are different topologies for PSO. Here the *Gbest* topology is used. It determines that all the particles are connected to each other and they are part of a unique neighborhood [65].

The general algorithm of PSO, with constriction factor, is shown in Algorithm 6.

3.5.2   Remarks on PSO

- PSO is able to find good enough solutions faster than other evolutionary algorithms [3].
- PSO has been hybridized with other optimization methods with success. The interest of combining PSO with other methods is mostly based on the fact that PSO does not have the ability of improving the quality of the solutions as the number of iterations is increased [3].

3.6   New Metaheuristic for Optimization: Differential Evolution with Particle Collision (DEwPC)

In this section the new metaheuristic *Differential Evolution with Particle Collision*, DEwPC, is described. Its objective is to improve the performance of DE based on the incorporation of some ideas from PCA. The idea for this algorithm emerged after the analysis of the results of the experiments with the Two Tanks system, see Chap. 4.

The new algorithm keeps the same structure of operators *Mutation* and *Crossover* from DE, while introduces a modification in the *Selection* operator.

Comparing with the traditional version of DE, DEwPC improves the *Selection* with the incorporation of the mechanism *Absorption-Scattering* of PCA. This new selection only increases in one of the number of parameters of DE. The added parameter is $MaxIter_c$, which comes from the *Local* operator in PCA.

The new *Selection* also provides DE with a simple *Metropolis* strategy. This allows to balance the levels of intensification and diversification of the search. As a consequence, DEwPC is better than DE in case of search in more complicated spaces, or when dealing with very noisy data.

From the point of view of PCA, DEwPC provides PCA with the structure of a population-based method. DEwPC also improves the stochastic generation of feasible solutions, by means of operators *Mutation* and *Crossover* from DE. These operators provide DEwPC with a higher *memory* than in PCA.

3.6.1 Description of the DEwPC Algorithm

DEwPC has the same operators *Mutation* and *Crossover* as the algorithm DE (see descriptions in Sect. 3.2), with the generation mechanism $DE/X^{best}/2/bin$.

The new selection operator was denominated *Selection with probability*. It applies the *Absorption-Scattering* mechanism (see algorithm in Algorithm 2) from PCA (see Sect. 3.3).

The application of the new operator *Selection with probability* to all the Z possible new members X^z_{Iter} of the population could be very expensive computationally. In the worst case DEwPC could need $MaxIter \times Z \times MaxIter_c$ evaluations of the objective function; while DE needs $MaxIter \times Z$; and PCA needs $MaxIter \times MaxIter_c$. Considering $MaxIter = Z = MaxIter_c = n$, the complexity of the considered algorithms are: $O(n^2)$ for DE and PCA; and $O(n^3)$ for DEwPC. Therefore, it was decided to execute the new selection operator only to a certain number of members of the population: integer part of $\sqrt{Z}$. For the rest of the members, DEwPC applies the operator *Selection* of DE in its original version. This reduces the complexity of DEwPC to $O(n^{\frac{5}{2}})$.

The algorithm for DEwPC is presented in Algorithm 7.

3.6.2 Remarks on DEwPC

- DEwPC was inspired by the results of the comparison of the application of DE and PCA to diagnosing the benchmark *Two Tanks* (see Sect. 4.4).
- The objective of DEwPC is to improve the ability of DE to escape from local optima), based on the incorporation of some ideas from PCA.

Algorithm 7: Algorithm for Differential Evolution with Particle Collision (DEwPC)

Data: $C_{cross}, C_{scal}, Z, MaxIter, MaxIter_c$
Result: Best solution: X^{best}
Generate an initial random population of Z solutions;
Select best solution X^{best};
for $l = 1$ *to* $l = MaxIter$ **do**

> Apply *Mutation*, Eq. (3.10);
> Apply *Crossover*, Eq. (3.11);
> For $\hat{X}_l^{(z)}$, with $z \leq \sqrt{Z}$, apply *Selection* from DE, Eq. (3.12);
> For $\hat{X}_l^{(z)}$, with $z > \sqrt{Z}$, apply *Absorption-Scattering*($\hat{X}_l, X^{best}, X_{l-1}, MaxIter_c$) from PCA (see algorithm in Algorithm 2);
> Update X^{best};
> Verify stopping criteria;

end

- The operators *Mutation* and *Crossover* from DE provide DEwPC with an easy structure and easy implementation.
- In DEwPC the operator *Selection* from DE was modified. Most variations of DE are focused on modifications in *Mutation* and in the self-adaptation of its parameters [15, 27, 91, 102, 129, 144, 146, 148]. This is an important difference between this proposal and the well-known variations of DE.
- The modification in the *Selection* from DE adds only one new parameter to DEwPC with respect to DE.

3.6.3 Validation of DEwPC

In order to validate DEwPC for continuous optimization problems without constrains, five test functions have been considered. They are within the suite of test functions recommended in the literature [30, 45, 125]:

- F1: Shifted Sphere Function.
- F2: Shifted Schwefel's Problem 1.2 (Let's remark that here Problem 1.2 is referred to the name of this function and not to a problem described in this book).
- F4: Shifted Schwefel's Problem 1.2 with noise in fitness.
- F6: Shifted Rosenbrock's Function.
- F7: Shifted Rotated Griewank's Function without bounds.

In References [30, 45, 125] the readers can find the description of the mentioned test functions.

The experiments and evaluation criteria follow the indications recommended in [45, 125]. The experiments consider ($\mathbb{D} \subset \mathbb{R}^n$) first $n = 10$ and later $n = 30$. Functions F1, F2, and F4 are unimodal; F6 is multimodal for the dimensions used in the experiments ($n = 10$ and $n = 30$); and F7 is also multimodal. The description of the experiments is as follows:

- **Experiment A:** The stopping criteria are the maximum number of objective function evaluations, $Eval_{max} = 10,000n$, or the maximum error $ErrTer = 10^{-8}$. The latter is a convergence tolerance. The following quantities are separately computed for each problem:

  - Average and standard deviation of the number of evaluations of the objective function and error.
  - Success Rate, $SR = \frac{EE}{ET}$: being EE the number of successful runs, and ET the number of total runs.
  - Success Performance, $SP = \frac{\overline{Eval}_{EE}}{SR}$, where $\overline{Eval}_{EE}$ is the average of the number of function evaluations for successful runs.

  A successful run is understood as a one in which the algorithm achieves a fixed accuracy level, $ErrTer$, within the prescribed maximum number of function evaluations, $Eval_{max}$.
- **Experiment B:** It was considered $Eval_{max}$ as the only one stopping criterion. Different values were taken: $Eval_{max} = 100n$, $1000n$, and $10,000n$, respectively. The Final error, $Error_f$, was determined for each case:

$$Error_f = f^* - f^{best} \tag{3.34}$$

being f^* the optimal value of the function and f^{best} the best value of the function that the algorithm provided.

For each test function, the algorithms were run 25 times, for each experiment. This number allows to reach statistically valid conclusions, concerning the performance of the algorithms [30, 45, 125]. The computer used during all the experiments has a processor Inter(R) Core(TM) i7-4500U CPU @ 1.80 GHz 2.40 GHz and a RAM of 8 GB.

In order to suggest values for the parameters of DEwPC, some tests with different sets of values for Z and $MaxIter_c$ were made, see Table 3.1. With the aim of decreasing the computational cost of DEwPC (with respect to DE), the size of the population Z is decreased; and $MaxIter_c$ is made dependent on Z (which depends on the size of the problem $dim\mathbb{D} = n$).

In Table 3.1 the different sets of values are shown. In all cases, the constants parameters $C_{cross} = 0.9$ and $C_{scal} = 0.5$ were considered as recommended in [111, 124]. The notation [.] indicates the nearest integer to the value inside.

For function F6, it is well known that the global optimum is inside a long, narrow, parabolic shaped valley. Therefore, the convergence to the global optimum is difficult. Hence, this problem is often used in assessing the performance of optimization algorithms. Therefore, this function was used for deciding the set of parameter values to be used in the further implementations of DEwPC.

In Fig. 3.2 it is shown a comparison of results for F6, with $n = 10$ and $n = 30$, and the sets of values shown in Table 3.1. In all cases, $Eval_{max} = 10,000n$ was considered as the stopping criterion. The better results are for Cases 4, 5, and 6, which have half of recommended population in DE. The best result is for $MaxIter_c = 0.2Z$, i.e. Set 5 in Table 3.1.

Table 3.1 Sets of values for parameters of DEwPC

Set	Z	$MaxIter_c$
1	10n	[0.3Z]
2	10n	[0.2Z]
3	10n	[0.1Z]
4	5n	[0.3Z]
5	5n	[0.2Z]
6	5n	[0.1Z]
7	2n	[0.3Z]
8	2n	[0.2Z]
9	2n	[0.1Z]

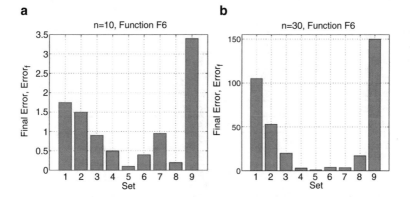

Fig. 3.2 Final error Er_f obtained with DEwPC in the minimization of F6, $n = 10$ (**a**) and $n = 30$ (**b**)

Table 3.2 Recommended parameter values for DEwPC

Alg	Z	C_{scal}	C_{cross}	$MaxIter_c$
DEwPC	5n	0.5	0.9	[0.2Z]

Therefore, it was used the Set 5 during Experiments A and B, see Table 3.2. These are then considered as the recommended values for the parameters of DEwPC.

3.6.3.1 Results for Experiment A with DEwPC

In Tables 3.3 and 3.4 are shown the results for the test cases with $n = 10$. Table 3.3 shows the information related with the final error. The error values for the most representative runs are presented: the best; the worst; and runs in the 7th, 13th, and 19th position (the runs are in ascending order considering the final error as the ordering criterion). The values for the error average and its standard deviation for each algorithm are also presented. Table 3.4 shows similar information but regarding the number of evaluations of the objective function.

Table 3.3 Error analysis for Experiment A with DEwPC and test cases with dimension $n = 10$

Run (ordered)	F1-DE	F1-DEwPC	F2-DE	F2-DEwPC
1(best)	$0.2736 \cdot 10^{-8}$	$0.3217 \cdot 10^{-8}$	$0.4254 \cdot 10^{-8}$	$0.4302 \cdot 10^{-8}$
7	$0.5682 \cdot 10^{-8}$	$0.7789 \cdot 10^{-8}$	$0.7419 \cdot 10^{-8}$	$0.7175 \cdot 10^{-8}$
13 (median)	$0.7975 \cdot 10^{-8}$	$0.8443 \cdot 10^{-8}$	$0.8257 \cdot 10^{-8}$	$0.7737 \cdot 10^{-8}$
19	$0.9488 \cdot 10^{-8}$	$0.9316 \cdot 10^{-8}$	$0.9364 \cdot 10^{-8}$	$0.8731 \cdot 10^{-8}$
25(worst)	$0.9817 \cdot 10^{-8}$	$0.9885 \cdot 10^{-8}$	$0.9984 \cdot 10^{-8}$	$0.9977 \cdot 10^{-8}$
Average	$0.74713 \cdot 10^{-8}$	$0.81668 \cdot 10^{-8}$	$0.81417 \cdot 10^{-8}$	$0.75799 \cdot 10^{-8}$
Standard deviation	$0.22110 \cdot 10^{-8}$	$0.15113 \cdot 10^{-8}$	$0.14600 \cdot 10^{-8}$	$0.15254 \cdot 10^{-8}$
Run (ordered)	F4-DE	F4-DEwPC	F6-DE	F6-DEwPC
1(best)	$0.4809 \cdot 10^{-8}$	$0.4542 \cdot 10^{-8}$	$0.1 10^{-8}$	$0.1 10^{-8}$
7	$0.7892 \cdot 10^{-8}$	$0.7944 \cdot 10^{-8}$	$0.1 10^{-8}$	$0.1 10^{-8}$
13(median)	$0.8753 \cdot 10^{-8}$	$0.8915 \cdot 10^{-8}$	$0.1 10^{-8}$	$0.1 10^{-8}$
19	$0.9138 \cdot 10^{-8}$	$0.9506 \cdot 10^{-8}$	$0.1 10^{-8}$	$0.1 10^{-8}$
25(worst)	$0.9706 \cdot 10^{-8}$	$0.9933 \cdot 10^{-8}$	3.9866	4.506
Average	$0.81994 \cdot 10^{-8}$	$0.84913 \cdot 10^{-8}$	0.3189	0.6379
Standard deviation	$0.14243 \cdot 10^{-8}$	$0.12822 \cdot 10^{-8}$	1.1038	1.4916
Run (ordered)	F7-DE	F7-DEwPC		
1(best)	0.0616	0.0172		
7	0.1304	0.0369		
13(median)	0.2978	0.0590		
19	0.3855	0.0786		
25(worst)	0.5067	0.1402		
Average	0.2737	0.0643		
Standard deviation	0.1475	0.0346		

In Table 3.3 it is shown that both algorithms, DE and DEwPC, reached the desired error, i.e Error$_f <$ $ErrTer = 10^{-8}$, for functions F1, F2, and F4. Table 3.4 shows that for these cases, the best value for the average $\overline{Eval}$ and SP were obtained with DEwPC. It can be concluded that DEwPC improves the accuracy of estimations for unimodal functions (F1 and F2), and also with noisy environments (F4). For function F6 not all the runs were successful and the SP of DEwPC is better, i.e. smaller value for SP. Table 3.4 also shows that for function F6 the SR achieved by DEwPC is better than the one with DE, i.e. higher value for SR. This is an indication that the new Selection operator provides DEwPC with a better local search capability and better searching diversity. DE and DEwPC did not reach the desired error for F7, i.e. $ErrTer = 10^{-8}$ but, as shown in Table 3.3, the error that DEwPC reached for the 100,000 evaluations of the objective functions, was smaller.

The results for the test cases with $n = 30$ are shown in Tables 3.5 and 3.6. They show that for $n = 30$ the algorithms only reached the desired error in all the runs for functions F1 and F2. In these cases, the values of SP from DEwPC are better, i.e. smaller values. It means that for unimodal functions DEwPC leads to a reduction on the computational cost. DE did not have successful runs for F4, F6, and F7,

Table 3.4 Analysis of the number of evaluations of the objective function for Experiment A with DEwPC and test cases with dimension $n = 10$

Run (ordered)	F1-DE	F1-DEwPC	F2-DE	F2-DEwPC
1(best)	21,900	15,530	23,000	18,040
7	22,700	16,090	23,800	18,860
13(median)	23,300	16,200	24,300	19,330
19	23,600	16,840	24,800	19,920
25(worst)	24,400	17,980	25,400	20,720
$\overline{Eval}$	23,148	16,431	24,312	19,363
Standard deviation	651.3	576.8	656.5	764.2
EE	25	25	25	25
SR%	100	100	100	100
SP	23,148	16,431	24,312	19,363
Run (ordered)	F4-DE	F4-DEwPC	F6-DE	F6-DEwPC
1(best)	40,200	29,870	45,500	39,080
7	41,700	30,620	47,400	47,990
13(median)	42,500	31,710	49,000	49,010
19	43,300	32,680	56,700	66,000
25(worst)	45,900	34,760	100,000	100,000
$\overline{Eval}$	42,700	31,826	57,800	50,767
Standard deviation	1570.6	1403.8	16,417	19,761
EE	25	25	23	21
SR%	100	100	92	84
SP	42,700	31,826	59,828	61,031
Run (ordered)	F7-DE	F7-DEwPC		
1(best)	100,000	100,000		
7	100,000	100,000		
13(median)	100,000	100,000		
19	100,000	100,000		
25(worst)	100,000	100,000		
$\overline{Eval}$	100,000	100,000		
Standard deviation	0	0		
EE	0	0		
SR%	0	0		
SP	–	–		

while DEwPC was able to obtain successful runs for F4 and F7. Therefore, it can be concluded that for more complex search spaces DEwPC offers a better performance than DE.

Table 3.7 shows a summary of the evaluation criteria, as well as the normalized SP with respect to the best, SP_{best}, between both algorithms; and the index δSP, which indicates the percentage increase in SP when going from a space with $n = 10$ to $n = 30$. In case, that some of the indicators cannot be computed due to lack of successful runs, it is indicated by means of a bar $-$, and the average of the final error

Table 3.5 Error analysis for Experiment A with DEwPC and test cases with dimension $n = 30$

Run (ordered)	F1-DE	F1-DEwPC	F2-DE	F2-DEwPC
1(best)	$0.8534 \cdot 10^{-8}$	$0.7026 \cdot 10^{-8}$	$0.3779 \cdot 10^{-8}$	$0.3103 \cdot 10^{-8}$
7	$0.8978 \cdot 10^{-8}$	$0.8290 \cdot 10^{-8}$	$0.5973 \cdot 10^{-8}$	$0.8166 \cdot 10^{-8}$
13(median)	$0.9277 \cdot 10^{-8}$	$0.9046 \cdot 10^{-8}$	$0.8233 \cdot 10^{-8}$	$0.8792 \cdot 10^{-8}$
19	$0.9497 \cdot 10^{-8}$	$0.9320 \cdot 10^{-8}$	$0.8908 \cdot 10^{-8}$	$0.9653 \cdot 10^{-8}$
25(worst)	$0.9817 \cdot 10^{-8}$	$0.9924 \cdot 10^{-8}$	$0.9923 \cdot 10^{-8}$	$0.9942 \cdot 10^{-8}$
Average	$0.92594 \cdot 10^{-8}$	$0.87839 \cdot 10^{-8}$	$0.74867 \cdot 10^{-8}$	$0.83785 \cdot 10^{-8}$
Standard deviation	$0.04.1291 \cdot 10^{-8}$	$0.083257 \cdot 10^{-8}$	$0.17459 \cdot 10^{-8}$	$0.17361 \cdot 10^{-8}$
Run (ordered)	F4-DE	F4-DEwPC	F6-DE	F6-DEwPC
1(best)	299.50	$2.12 \cdot 10^{-9}$	71.50	3.263
7	362.70	0.180	131.41	16.89
13(median)	0.426	236.44	184.85	18.69
19	495.21	0.267	299.784	19.56
25(worst)	687.64	0.406	553.90	76.68
Average	441.3	0.228	300.01	25.87
Standard deviation	110.1	0.87	143.4	20.8
Run (ordered)	F7-DE	F7-DEwPC		
1(best)	0.0025	10^{-8}		
7	0.0085	10^{-8}		
13(median)	0.0220	10^{-8}		
19	0.0673	0.0099		
25(worst)	0.7435	0.0222		
Average	0.1096	0.0051		
Standard deviation	0.1931	0.0065		

is put inside brackets. The bold values represent the best values obtained between DE and DEwPC.

Table 3.7 shows that DEwPC, for $n = 10$, has the best SP in four out of the five test functions (F1, F2, F4, and F6). For F7 both algorithms did not achieve successful runs, but the final error with DE was higher than the final error with DEwPC. Table 3.7 also shows that for function F6, with $n = 10$, the SR achieved by DEwPC was better than the SR obtained with DE (see Fig. 3.3 for a graphical representation of these results). This indicates that the new Selection operator provides DEwPC with a better local search capability and better searching diversity. Table 3.7 also shows that for the test cases with $n = 30$, DEwPC achieved successful runs in also four out of the five test functions (F1, F2, F4 and F7), while DE did it only twice (F1 and F2). These results indicate that in more complex search spaces (with higher dimension), DEwPC allows to reach better results than DE. In other words, the modification in the Selection operator provides DEwPC with a more effective way to avoid being in local minima.

It was also applied the Wilcoxon's Test in order to reach a conclusion on the differences between the SP values. For this purpose, it was used the function *signrank* from Matlab®. The results showed, that the SP values obtained with DEwPC are better with p-value $p = 0.013$.

Table 3.6 Analysis of the number of evaluations of the objective function for Experiment A with DEwPC and test cases with dimension $n = 30$

Run(ordered)	F1-DE	F1-DEwPC	F2-DE	F2-DEwPC
1(best)	234,090	163,500	111,600	103,890
7	249,240	169,800	113,400	109,050
13(median)	253,020	173,100	114,300	114,030
19	255,270	174,740	116,400	116,100
25(worst)	271,830	175,200	119,400	122,100
$\overline{Eval}$	252,560	172,752	114,744	113,600
Standard deviation	7723.0	4550.3	1975.1	4873.8
EE	25	25	25	25
SR%	100	100	100	100
SP	252,560	172,752	114,744	113,600
Run (ordered)	F4-DE	F4-DEwPC	F6-DE	F6-DEwPC
1(best)	300,000	256,200	300,000	300,000
7	300,000	300,000	300,000	300,000
13(median)	300,000	300,000	300,000	300,000
19	300,000	300,000	300,000	300,000
25(worst)	300,000	300,000	300,000	300,000
$\overline{Eval}$	300,000	298,248	300,000	300,000
Standard deviation	0	8760	0	0
EE	0	1	0	0
SR%	0	4	0	0
SP	–	6,405,000	–	–
Run (ordered)	F7-DE	F7-DEwPC		
1(best)	300,000	243,100		
7	300,000	255,970		
13(median)	300,000	272,340		
19	300,000	300,000		
25(worst)	300,000	300,000		
$\overline{Eval}$	300,000	276,640		
Standard deviation	0	22,858		
EE	0	14		
SR%	0	56		
SP	–	460,960		

3.6.3.2 Results for Experiment B with DEwPC

In Tables 3.8 and 3.9 are presented the average values of the final errors from Experiment B. Let's note that the bold values in Tables 3.8 and 3.9, represent the best values obtained between PSO-M and PSO. In particular, Table 3.8 shows that for F1, F2, F4 and test cases with $n = 10$, DEwPC always provides a smaller error. For F6 and F7, this does not occur.

Table 3.7 Summary of the evaluation criteria for Experiment A with DEwPC

Alg	Function	$n = 10$			$n = 30$			$\delta SP \cdot 100\%$
		SR	SP	SP/SP$_{best}$	SR	SP	SP/SP$_{best}$	
DE	F1	100	23,148	1.41	100	252,560	1.46	90.8
DEwPC	F1	100	**16,431**	1	100	**172,752**	1	90.5
DE	F2	100	24,312	1.26	100	114,744	1.01	78.8
DEwPC	F2	100	**19,363**	1	100	**113,600**	1	82.9
DE	F4	100	42,700	1.34	0	–	[441.3151]	–
DEwPC	F4	100	**31,826**	1	4	**6,405,000**	1	99.5
DE	F6	84	61,031	1.02	0	–	[300.01]	–
DEwPC	F6	92	**59,828**	1	0	–	[25.8751]	–
DE	F7	0	–	[0.2737]	0	–	[0.1096]	–
DEwPC	F7	0	–	**[0.0643]**	56	**460,960**	1	–

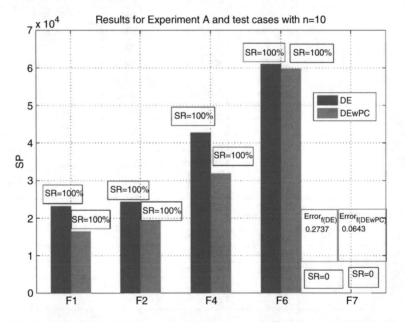

Fig. 3.3 Summary of the evaluation criteria: SP, SR and error for Experiment A with DEwPC, and test cases with dimension $n = 10$

Table 3.9 shows that for F1 and F2, DEwPC reached 30,000 evaluations with a smaller error. For F4, F6, and F7, DEwPC always had the best result.

In Fig. 3.4 it is represented the value of the ratio between the average error obtained with DE and DEwPC, for each function, for the test cases with $n = 30$. The average error for DEwPC is smaller than the average error for DE. For F4 and F7, a big difference keeps until the final number of evaluations.

In Fig. 3.5 it is graphically shown the results for Experiment B, and the test functions F4, F6, and F7, for the test cases with $n = 10$ and $n = 30$. It can be

Table 3.8 Final error analysis for Experiment B with DEwPC, and test cases with dimension $n = 10$

Alg	Function	Error $f(Eval_{max}=1000)$	Error $f(Eval_{max}=10,000)$	Error $f(Eval_{max}=100,000)$
DE	F1	65.73	0.0604	0
DEwPC	F1	**36.13**	**$2.490 \cdot 10^{-5}$**	0
DE	F2	59.97	1.1252	0
DEwPC	F2	**57.04**	**0.0017**	0
DE	F4	102.95	6.8093	0
DEwPC	F4	**18.83**	**0.0045**	0
DE	F6	$7.646 \cdot 10^4$	41.260	0.63
DEwPC	F6	**$2.091 \cdot 10^4$**	**5.995**	**0.32**
DE	F7	**0.7039**	0.6347	0.2181
DEwPC	F7	0.9223	**0.3735**	**0.07**

Table 3.9 Final error analysis for Experiment B with DEwPC, and test cases with dimension $n = 30$

Alg	Function	Error $f(Eval_{max}=3000)$	Error $f(Eval_{max}=30,000)$	Error $f(Eval_{max}=300,000)$
DE	F1	613.4	257.7	0
DEwPC	F1	**461.0**	**22.09**	0
DE	F2	**36.92**	1.117	0
DEwPC	F2	52.64	**0.032**	0
DE	F4	$1.071 \cdot 10^3$	863.7	431.3151
DEwPC	F4	**867.6**	**167.2**	**0.237**
DE	F6	$2.362 \cdot 10^6$	$1.164 \cdot 10^6$	303.0
DEwPC	F6	**$8.295 \cdot 10^5$**	**$1.585 \cdot 10^3$**	**18.60**
DE	F7	**1.119**	1.055	0.539
DEwPC	F7	1.134	**0.936**	**0.017**

observed that the final error for DEwPC is in almost all cases lower than the final error for DE, no matter the dimension n on the problem considered.

Furthermore, the Wilcoxon's test showed that the error that achieves DEwPC, when the algorithm has executed 10% of the maximum number of objective function evaluations, is lower than the error achieved with DE, under the same conditions, with p-value $p = 0.0020$. For this analysis, it was used the function *signrank* from Matlab®.

3.7 New Metaheuristic for Optimization: Particle Swarm Optimization with Memory (PSO-M)

In this section the new metaheuristic *Particle Swarm Optimization with Memory* (PSO-M) is presented. PSO-M is inspired on the metaheuristics PSO and ACO, that were described in Sects. 3.5 and 3.4, respectively.

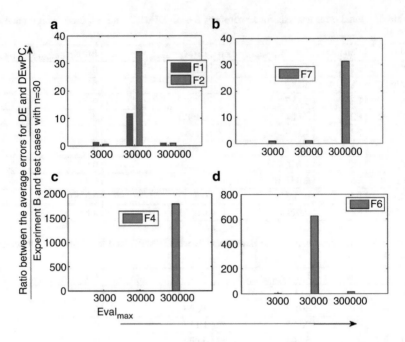

Fig. 3.4 Ratio between the average error obtained with DE and DEwPC for Experiment B, and test cases with dimension $n = 30$

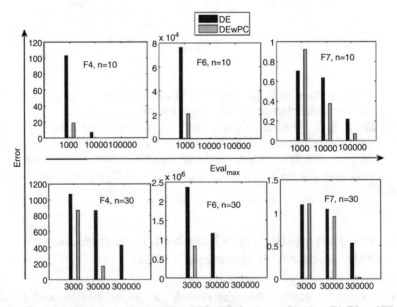

Fig. 3.5 Error from Experiment B with DE and DEwPC; for the test functions F4, F6, and F7; and for the test cases with dimensions $n = 10$ and $n = 30$

PSO-M has the objective of reducing the number of iterations/generations that PSO needs to execute, in order to obtain reasonably good quality solutions. For that, the new algorithm adds to PSO, in a simple way, a learning mechanism. That allows to use previous information in a better way and, as a consequence, the search is upgraded. PSO-M incorporates the pheromone matrix of ACO for storing the historical behaviors (memory) that are addressed in the future search directions.

The algorithm has two stages:

- **First stage:** A swarm explores the search space, i.e. canonical PSO is applied. A pheromone matrix stores the information of the exploration performed.
- **Second stage:** Another swarm makes an intensification on the promising regions of the search space. For that purpose, its initial position is generated using the generation scheme of ACO with the pheromone matrix achieved in the first stage.

3.7.1   Description of the PSO-M Algorithm

In the first stage, a swarm of Z_1 agents explores the search space $\mathbb{D}$ following the canonical structure of PSO. The new position $X^z_{Iter_1} \in \mathbb{R}^n$ and the new velocity $V^z_{Iter_1}$ of each particle z, with $z = 1, 2, \ldots Z_1$, are obtained from Eqs. (3.27) and (3.32), respectively.

The values for PSO's parameters in this stage are the ones recommended in the literature [69, 111], except for Z_1. For the size of the swarm, its value is reduced to a half of the recommended. The maximum number of iterations in this stage is taken as 10% of the maximum number of iterations for the whole algorithm.

Following the idea of ACO for the continuous optimization case, the prescribed permissible interval for each decision variable x_i, with $i = 1, 1, \ldots, n$, is divided into K possible values x_i^k, with $k = 1, 2, \ldots, K$. The pheromone matrix $\mathbb{F} \in M_{n \times K}(\mathbb{R})$ is generated and updated at each iteration of this first stage of the PSO-M algorithm.

In order to update the pheromone matrix $\mathbb{F}$, the following strategy is proposed:

1. At each iteration $Iter_1$, each component of the vector X^{gbest} is identified with only one of the K discrete values, which are assigned to the variable, that corresponds to that component. This connection generates the vector $X^{gbest(d)}$. The components G_n of a vector function $G : \mathbb{R}^n \longmapsto \mathbb{R}^n$ are defined as:

Definition 3.2 Let x_i^{gbest} and $x_i^{gbest(d)}$ be the ith components of the vectors X^{gbest} and $X^{gbest(d)}$, respectively. Let G_i be the ith component function from the vector function G. The components G_i *of the vector function* G *are defined as:*

$$G_i(X^{gbest}) = x_i^{gbest(d)} = x_i^{\bar{m}} \tag{3.35}$$

where

$$\bar{m} : \left| x_i^{gbest} - x_i^{\bar{m}} \right| = \min_m \left| x_i^{gbest} - x_i^m \right| \text{ with } m = 1, 2 \ldots K \tag{3.36}$$

2. With the new vector $G(X^{gbest}) = X^{gbest(d)}$, matrix $\mathbb{F}$ is updated as in ACO, see Eq. (3.18).

The second stage of the PSO-M algorithm considers a swarm of Z_2 particles, with $Z_2 < Z_1$. This swarm performs an intensification of the promising regions of the search space $\mathbb{D}$, taking advantage of the information that was stored in matrix $\mathbb{F}$, from the first stage. For that purpose, the swarm is generated based on the mechanism from ACO using the pheromone matrix $\mathbb{F}$ from the first stage of PSO-M algorithm, see Eq. (3.20). With the parameter q_0, the level of confidence on the results from the first stage is weighted.

The inclusion of the second stage may increase the ability to escape from local optima to which the traditional PSO may converge. It also allows to make an intensification of the promising areas in the search space.

In Algorithm 8, the algorithm for the PSO-M metaheuristic is given.

Algorithm 8: Algorithm for Particle Swarm Optimization with Memory (PSO-M)

Data: $c_1, c_2, Z_1, Z_2, MaxIter_1, MaxIter_2, C_{evap}, C_{inc}, q_0, K$
Result: X^{best}
Discretize the domain of the n variables in K values;
Generate a random initial pheromone matrix $\mathbb{F}$;
Compute matrix PC with Eq. (3.17);
Generate randomly X_0^z and V_0^z for the Z_1 particles;
Update $X^{z(pbest)}$ and X^{gbest};
First Stage $(c_1, c_2, Z_1, MaxIter_1)$
for $l = 1$ *to* $l = MaxIter_1$ **do**
    for $z = 1$ *to* $z = Z_1$ **do**
        Update V_l^z with Eq. (3.32);
        Update X_l^z with Eq. (3.27);
        Update $X^{z(pbest)}$
    end
    Update X^{gbest};
    Compute $X^{gbest(d)}$ with Eq. (3.35);
    Update $\mathbb{F}$ with Eq. (3.18);
end
Second Stage $(\mathbb{F}, q_0, c_1, c_2, Z_2, MaxIter_2)$
Generate X_0^z, for the Z_2 particles with $\mathbb{F}$ and mechanism of ACO;
Execute PSO (see Algorithm 6), with parameters $(c_1, c_2, Z_2, MaxIter_2)$ and the following variation: only V_0^z are randomly generated, the X_0^z are from the previous line;
Solution: $X^{best} X^{gbest}$

Table 3.10 Recommended parameter values for PSO-M

First stage

Z_1	c_1	c_2	ω	$Eval1_{max}$	k	q_0	C_{inc}	C_{evap}
$15n$	2.05	2.05	0.729	$10\%Eval_{max}$	63	0.55	0.30	0.10

Second stage

Z_2	c_1	c_2	ω	$Eval2_{max}$				
$5n$	2.05	2.05	0.729	$90\%Eval_{max}$				

The idea behind PSO-M is to reduce the computational cost based on the introduction of a pheromone matrix, and not in variations on parameter values of PSO. As a consequence, in both stages of PSO-M, the values of the PSO's parameters coincide with the parameter values of the PSO most recommended in the literature ($c_1 = c_2 = 2.05$, $w = 0.729$). The values of the parameters related to the pheromone matrix, i.e. the parameters in PSO-M which comes from ACO, were taken as recommended in literature [113]. The recommended values for the parameters of PSO-M are summarized in Table 3.10.

3.7.2   Remarks on PSO-M

- PSO-M can be understood as the PSO with a learning mechanism. Other variants of PSO were also focused on this idea, e.g. [83, 147].
- PSO-M has an easy structure, simple implementation, and a lower computational cost than PSO.
- There are other works that also present two-stages algorithms based on ACO [48, 101].
- Other algorithms have also resulted from the hybridization of PSO and ACO. For example, in Ref. [110] it is proposed an algorithm in which ACO is used to perform a local search around each particle of PSO, at each iteration. In Ref. [72] birds and ants make a simultaneous search, and share the information at each iteration.

3.7.3   Validation of PSO-M

For the validation of PSO-M, the experiments described in Sect. 3.6.3 are replied. The comparison is made with the variant of PSO with the constriction coefficient and the recommended values in the literature ($c_1 = c_2 = 2.05$, $w = 0.729$). The parameter values for PSO-M coincide with the presented in Table 3.10.

Table 3.11 Error analysis for Experiment A with PSO-M, and test cases with dimension $n = 10$

Run(ordered)	F1-PSO	F1-PSO-M	F2-PSO	F2-PSO-M
1(best)	$0.4724 \cdot 10^{-8}$	$0.6015 \cdot 10^{-8}$	0.0011	10^{-9}
7	$0.7270 \cdot 10^{-8}$	$0.7274 \cdot 10^{-8}$	0.0099	10^{-9}
13(median)	$0.8645 \cdot 10^{-8}$	$0.8342 \cdot 10^{-8}$	0.0294	0.0001
19	$0.9530 \cdot 10^{-8}$	$0.9568 \cdot 10^{-8}$	0.0668	0.0003
25(worst)	$0.9930 \cdot 10^{-8}$	$0.9992 \cdot 10^{-8}$	1.1736	0.0027
Average	$0.82705 \cdot 10^{-8}$	$0.83371 \cdot 10^{-8}$	0.0999	$3.1591 \cdot 10^{-4}$
Standard deviation	$0.15326 \cdot 10^{-8}$	$0.12733 \cdot 10^{-8}$	0.2351	$5.5881 \cdot 10^{-4}$
Run(ordered)	F4-PSO	F4-PSO-M	F6-PSO	F6-PSO-M
1(best)	0.0131	0.0006	0.01	0.0450
7	0.3188	0.0217	5.5	2.2675
13(median)	0.5504	0.0571	6.6	4.7256
19	1.1457	0.3006	173.1	5.1776
25(worst)	15.2542	3.7647	5434.3	291.1
Average	1.6974	0.3631	476.2	21.97
Standard deviation	3.3522	0.7816	1237.8	59.89
Run(ordered)	F7-PSO	F7-PSO-M		
1(best)	0.0099	0.0172		
7	0.0492	0.0590		
13(median)	0.0713	0.0762		
19	0.1033	0.1009		
25(worst)	0.1920	0.1305		
Average	0.0791	0.0770		
Standard deviation	0.0438	0.0308		

3.7.3.1 Results for Experiment A with PSO-M

Tables 3.11 and 3.12 show the results for the test cases with $n = 10$. The notations used in these tables are the same as those described in Sect. 3.6.3. From the results in Tables 3.11 and 3.12 it is observed that both algorithms always reached the desired error on the estimations for the test function F1, being the value for SP obtained with PSO-M better. For the test function F2, only PSO-M had successful runs, with SR $= 32\%$. For the more complex test functions F4, F6, and F7, the algorithms did not reach the desired error, but for these three functions, the average errors on estimations obtained with PSO were higher than the ones obtained with PSO-M bigger: 4.72; 21.7; and 1.02 times higher, respectively. Therefore, PSO-M improves the performance of PSO in noisy environments and more complex search spaces.

Tables 3.13 and 3.14 show the results for the test cases with dimension $n = 30$. In Table 3.13 it is observed that the desired error on the estimations, i.e. Error$_f <$ $ErrTer = 10^{-8}$, was reached only once: with PSO-M for test function $F1$ and in Table 3.14 it is observed that this desired error was obtained with SR $= 64\%$. For the test functions F4, F6, and F7, the average error on the estimations obtained with

Table 3.12 Analysis of the number of evaluations of the objective function for Experiment A with PSO-M, and test cases with dimension $n = 10$

Run (ordered)	F1-PSO	F1-PSO-M	F2-PSO	F2-PSO-M
1(best)	86,000	80,500	100,000	70,000
7	88,500	82,450	100,000	70,050
13(median)	83,100	82,900	100,000	100,000
19	90,600	84,150	100,000	100,000
25(worst)	93,700	85,700	100,000	100,000
$\overline{Eval}$	89,668	83,326	100,000	88,960
Standard deviation	1841.8	1456.2	0	2094
EE	25	25	0	8
SR%	100	100	0	32
SP	89,668	83,326	–	218,750
Run (ordered)	F4-PSO	F4-PSO-M	F6-PSO	F6-PSO-M
1(best)	100,000	100,000	100,000	100,000
7	100,000	100,000	100,000	100,000
13(median)	100,000	100,000	100,000	100,000
19	100,000	100,000	100,000	100,000
25(worst)	100,000	100,000	100,000	100,000
$\overline{Eval}$	100,000	100,000	100,000	100,000
Standard deviation	0	0	0	0
EE	0	0	0	0
SR%	0	0	0	0
SP	–	–	–	–
Run (ordered)	F7-PSO	F7-PSO-M		
1(best)	100,000	100,000		
7	100,000	100,000		
13(median)	100,000	100,000		
19	100,000	100,000		
25(worst)	100,000	100,000		
$\overline{Eval}$	100,000	100,000		
Standard deviation	0	0		
EE	0	0		
SR%	0	0		
SP	–	–		

PSO was higher than the ones obtained with PSO-M: 5.69; 6.59; and 1.55 times higher, respectively. From Table 3.13 it is observed that for the test function F2, the estimations obtained with PSO were better.

Table 3.15 shows a summary with the results obtained for Experiment A. The notation is the same as the one introduced in Sect. 3.6.3 and the bold values in represent the best values obtained between PSO-M and PSO. PSO reached successful runs in one of the ten situations considered, i.e. five test functions F1, F2, F4, F6, and F7, and problems with two different dimensions: $n = 10$ and $n = 30$,

Table 3.13 Error analysis for Experiment A with PSO-M, and test cases with $n = 30$

Run(ordered)	F1-PSO	F1-PSO-M	F2-PSO	F2-PSO-M
1(best)	0.0001	$0.83 \cdot 10^{-8}$	8121	7548
7	0.0005	$0.96 \cdot 10^{-8}$	15,680	13,195
13(median)	0.001	$0.98 \cdot 10^{-8}$	14,474	23,574
19	0.0018	$2.31 \cdot 10^{-8}$	17,963	34,364
25(worst)	0.0074	$2.038 \cdot 10^{-5}$	25,651	69,773
Average	0.0016	$2.4605 \cdot 10^{-8}$	14,911	30,144
Standard deviation	0.0018	$4.0366 \cdot 10^{-8}$	4973.1	20,124
Run(ordered)	F4-PSO	F4-PSO-M	F6-PSO	F6-PSO-M
1(best)	13,369	1923	300	30
7	20,532	3925	2300	380
13(median)	29,822	5165	10,700	920
19	32,400	6565	31,000	3440
25(worst)	48,820	10,736	475,970	424,890
Average	29,977	5267	21,608	3276
Standard deviation	9770.3	2109	947,390	990,34
Run(ordered)	F7-PSO	F7-PSO-M		
1(best)	0.0172	0.0074		
7	0.0567	0.0418		
13(median)	0.0887	0.0713		
19	0.0994	0.0824		
25(worst)	0.1657	0.1083		
Average	0.0981	0.0633		
Standard deviation	0.0312	0.0281		

while PSO-M was successful in three of them in 3 (always for unimodal functions). The metaheuristics PSO-M achieved success for function F1 in both dimensions, but its SR decreased from $n = 10$ to $n = 30$ in 36%. These results are graphically presented in Figs. 3.6 and 3.7.

Figure 3.6 shows a summary of the results for Experiment A with PSO-M and PSO, and test cases with dimension $n = 10$. It is observed in Fig. 3.6a that for the test function F1, both algorithms achieved success, but the SP obtained with PSO-M is better (smaller). Figure 3.6a also shows that for the test function F2, the algorithm PSO-M achieved success and the algorithm PSO did not reach successful runs. Figure 3.6b–d shows that for the test functions F4, F6, and F7, respectively, both algorithms PSO and PSO-M did not reach success, but the errors obtained with PSO-M are always smaller than the ones obtained with PSO.

Figure 3.7a–d shows a summary of the results for Experiment A with PSO-M and PSO, test cases with dimension $n = 30$ and functions F2, F4, F6, and F7, respectively. It is observed that both algorithms, PSO and PSO-M, did not reach

Table 3.14 Analysis of the number of evaluations of the objective function for Experiment A with PSO-M, and test cases with dimension $n = 30$

Run (ordered)	F1-PSO	F1-PSO-M	F2-PSO	F2-PSO-M
1(best)	300,000	293,100	300,000	300,000
7	300,000	295,050	300,000	300,000
13(median)	300,000	297,900	300,000	300,000
19	300,000	300,000	300,000	300,000
25(worst)	300,000	300,000	300,000	300,000
$\overline{Eval}$	300,000	297,336	300,000	300,000
Standard deviation	0	2564.8	0	0
EE	0	16	0	0
SR%	0	64	0	0
SP	–	462,250	–	–
Run (ordered)	F4-PSO	F4-PSO-M	F6-PSO	F6-PSO-M
1(best)	300,000	300,000	300,000	300,000
7	300,000	300,000	300,000	300,000
13(median)	300,000	300,000	300,000	300,000
19	300,000	300,000	300,000	300,000
25(worst)	300,000	300,000	300,000	300,000
$\overline{Eval}$	300,000	300,000	300,000	300,000
Standard deviation	0	0	0	0
EE	0	0	0	0
SR%	0	0	0	0
SP	–	–	–	–
Run (ordered)	F7-PSO	F7-PSO-M		
1(best)	300,000	300,000		
7	300,000	300,000		
13(median)	300,000	300,000		
19	300,000	300,000		
25(worst)	300,000	300,000		
$\overline{Eval}$	300,000	300,000		
Standard deviation	0	0		
EE	0	0		
SR%	0	0		
SP	–	–		

success, but the errors obtained with PSO-M are smaller than the ones obtained with PSO for three of the four test functions represented, i.e. F4, F6, and F7. For the test function F1, only PSO-M achieved success with SR = 64% and for that reason a graphic comparison of the errors between both algorithms is not presented.

Table 3.15 Summary of the evaluation criteria for Experiment A with PSO-M

Alg	Function	$n = 10$			$n = 30$			$\delta SP \cdot 100\%$
		SR	SP	SP/SP_{best}	SR	SP	SP/SP_{best}	
PSO	F1	100	89,668	1.08	0	–	[0.0016]	–
PSO-M	F1	100	**83,326**	1	64	462,250	1	81.97
PSO	F2	0	–	[0.0999]	0	–	**[14,911]**	–
PSO-M	F2	32	218,750	1	0	–	[30,144]	–
PSO	F4	0	–	[1.6974]	0	–	[29,977]	–
PSO-M	F4	0	–	**[0.3631]**	0	–	**[5267]**	–
PSO	F6	0	–	[476.2820]	0	–	[21,608]	–
PSO-M	F6	0	–	**[21.9764]**	0	–	**[3276]**	–
PSO	F7	0	–	[0.0791]	0	–	[0.0981]	–
PSO-M	F7	0	–	**[0.0770]**	0	–	**[0.0633]**	–

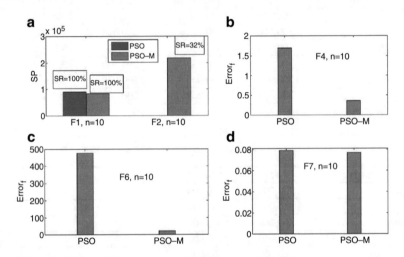

Fig. 3.6 Summary of evaluation criteria: SP, SR and error for Experiment A with PSO-M, and test cases with dimension $n = 10$

3.7.3.2 Results for Experiment B with PSO-M

Tables 3.16 and 3.17 show the behavior of the error on the estimations, obtained with both algorithms, PSO and PSO-M, for each test function, and test cases with $n = 10$ and $n = 30$, respectively. Let's note that the bold values in Tables 3.16 and 3.17, represent the best values obtained between PSO-M and PSO.

Table 3.16 shows that for both maximum number of function evaluations $Eval_{max} = 10,000$ and $Eval_{max} = 100,000$, PSO-M always achieved a smaller error. In Table 3.17 it is observed that the error averages obtained with PSO-M, for the different values of the maximum number of objective functions evaluations, are always smaller, except for the test function F2.

In Fig. 3.8 it is shown that the value of the ratio between the average errors obtained with PSO and PSO-M, for the test cases with $n = 30$, is greater than

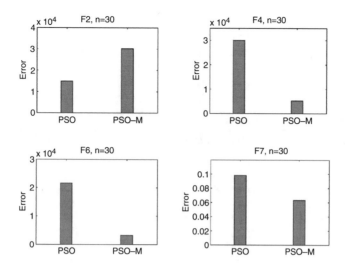

Fig. 3.7 Error for Experiment A with PSO-M, and test cases with $n = 30$

Table 3.16 Final error analysis for Experiment B with PSO-M, and test cases with dimension $n = 10$

Alg	Function	$\text{Error}_{Eval_{max}=1000}$	$\text{Error}_{Eval_{max}=10,000}$	$\text{Error}_{Eval_{max}=100,000}$
PSO	F1	$7.2640 \cdot 10^4$	$0.10714 \cdot 10^{-3}$	$0.22204 \cdot 10^{-15}$
PSO-M	F1	$\mathbf{1.7541 \cdot 10^4}$	**7.2868**	**0**
PSO	F2	$\mathbf{1.0526 \cdot 10^4}$	$1.023 \cdot 10^4$	0.099
PSO-M	F2	$1.9721 \cdot 10^4$	**5878.2**	$\mathbf{0.31591 \cdot 10^{-3}}$
PSO	F4	$1.0643 \cdot 10^4$	9193.3	1.6974
PSO-M	F4	**2611.0**	**35.809**	**0.3631**
PSO	F6	$2.7044 \cdot 10^9$	$4.8213 \cdot 10^8$	476.28
PSO-M	F6	$\mathbf{1.0336 \cdot 10^9}$	$\mathbf{3.4432 \cdot 10^5}$	**21.976**
PSO	F7	4.6760	2.8009	0.0791
PSO-M	F7	**3.2104**	**1.0481**	**0.0770**

Table 3.17 Final error analysis for Experiment B with PSO-M, and test cases with $n = 30$

Alg	Function	$\text{Error}_{Eval_{max}=3000}$	$\text{Error}_{Eval_{max}=30,000}$	$\text{Error}_{Eval_{max}=300,000}$
PSO	F1	$5.9806 \cdot 10^4$	$3.6615 \cdot 10^4$	0.0016
PSO-M	F1	$\mathbf{5.1804 \cdot 10^4}$	**5760.2**	$\mathbf{0.4037 \cdot 10^{-7}}$
PSO	F2	$1.0323 \cdot 10^5$	$8.0768 \cdot 10^4$	**14911**
PSO-M	F2	$1.1256 \cdot 10^5$	$1.2044 \cdot 10^5$	30144
PSO	F4	$1.2580 \cdot 10^5$	$8.1390 \cdot 10^4$	29977
PSO-M	F4	$\mathbf{1.1561 \cdot 10^5}$	$\mathbf{7.6299 \cdot 10^2}$	**5267**
PSO	F6	$5.8157 \cdot 10^{10}$	$1.5416 \cdot 10^{10}$	21,608
PSO-M	F6	$\mathbf{5.9771 \cdot 10^4}$	**1477.6**	**3276**
PSO	F7	**17.8861**	13.2519	0.0981
PSO-M	F7	$7.7348 \cdot 10^4$	**8.9843**	**0.0633**

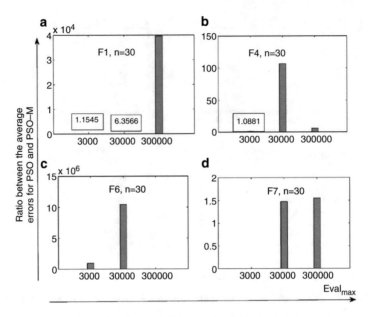

Fig. 3.8 Ratio between the average error for PSO and PSO-M; for Experiment B; test cases with dimension $n = 30$; and for test functions (a) F1, (b) F4, (c) F6 and (d) F7

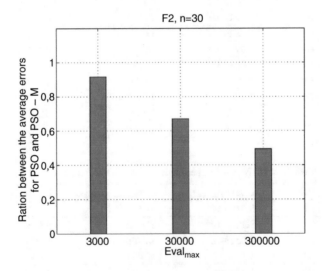

Fig. 3.9 Ratio between the average error for PSO and PSO-M; for Experiment B; test cases with dimension $n = 30$; and for test function F2

one, except for the test function F7, when the maximum of 3000 evaluations is considered. At the end of the 30,000 and 300,000 evaluations, the value of this ratio is always greater than one. This situation does not occur for function F2, see Fig. 3.9.

Furthermore, the Wilcoxon's test showed that the average error of PSO-M at $Eval_{max} = 10000n$ is lower than the error achieved by PSO, under the same conditions, with p-value $p = 0.0488$. For this analysis, it was used the function *signrank* from Matlab®.

3.8 Remarks

In this chapter was made an introduction to metaheuristics for optimization, as well as a classification of them. Description for the metaheuristics Differential Evolution; Particle Collision Algorithm; Ant Colony Optimization for continuous problems; and Particle Swarm Optimization was presented.

For the case of ACO-d, its adaptation and formalization to continuous optimization problems was also made and described in Sect. 3.4.

Two new metaheuristics for continuous optimization were described and validated: DEwPC and PSO-M, see Sects. 3.6 and 3.7, respectively. A hybrid strategy with ACO and DE was also presented in Sect. 3.4. These are powerful tools that will be used in Chap. 4 to Fault Diagnosis Inverse Problems for the benchmarks presented in Sect. 2.4.

Chapter 4
Applications of the Fault Diagnosis: Inverse Problem Methodology to Benchmark Problems

This chapter presents the application of the last three steps (2, 3, and 4) of the Fault Diagnosis—Inverse Problem Methodology (FD-IPM), as described in Sect. 2.1, to the three benchmark problems that were presented in Sect. 2.4. Step 2 is to evaluate the faults for which only the detection is possible. The third step deals with the solution of an optimization problem with metaheuristics. The fourth step provides the conclusion on the diagnosis of the system, based on the results from the third step. The experiments with the DC Motor, Inverted Pendulum System, and Two Tanks System are presented in Sects. 4.2, 4.3, and 4.4, respectively. The Inverted Pendulum System is affected by faults which cannot be diagnosed, only detected. For that reason, Step 2 was only applied to the the Inverted Pendulum System.

Table 4.1 shows the acronyms of the metaheuristics that were applied to each benchmark problem. Section 4.1 describes important aspects to be considered during the experiments. The experiments are designed in order to analyze robustness and sensitivity of the diagnosis obtained with FD-IPM.

4.1 Experimental Methodology

For diagnosing with the Fault Diagnosis—Inverse Problem Methodology (FD-IPM), the first step is to formulate the FDI problem as an Inverse Problem for each benchmark problem and checking the hypothesis H1–H3, presented in Sect. 2.1. That was made in Sect. 2.5 for each benchmark problem. Based on Step 1 of FD-IPM, it is possible to decide if the faults can only be detected (go to Step 2 of FD-IPM), diagnosed (go to Step 3 of FD-IPM), or cannot be detected with FD-IPM.

If the faults can only be detected, then the second step of the methodology must be applied. In this step the value of the objective function $f(\hat{\boldsymbol{\theta}}_f)$ of the optimization problem, which is formulated in Step 1 is computed and compared with a certain

© Springer International Publishing AG, part of Springer Nature 2019
L. Camps Echevarría et al., *Fault Diagnosis Inverse Problems: Solution with Metaheuristics*, Studies in Computational Intelligence 763,
https://doi.org/10.1007/978-3-319-89978-7_4

Table 4.1 Algorithms
applied to each benchmark
problem

Benchmark	Algorithms
DC Motor	ACO, PSO, PSO-M
IPS	ACO, ACO-d
Two Tanks System	ACO, DE, PCA, DEwPC, ACO-DE

threshold value, f_{umbral}. As described in Chap. 2, the value of f_{umbral} is important in order to have a good balance between robustness and sensitivity of the detection result. The application of Step 2 for detecting the faults in the Inverted Pendulum System is used to show how to determine this value.

If the faults can be diagnosed, then the third step of FD-IPM must be applied in order to estimate the fault values. Step 3 of FD-IPM deals with the solution of the optimization problem, see the definition of the optimization problem given in Eq. (2.3), which is solved with the use of metaheuristics. Step 4 of the methodology provides the result of the diagnosis: If any component of $\hat{\theta}_f$, which is obtained from Step 3, is different from zero, then the fault that corresponds to that component is affecting the system. The magnitude of the faults coincides with the estimated obtained in Step 3. Therefore, the diagnosis is concluded when the estimated values for the faults are obtained.

It is clear that Step 3 is of major importance: the final result of the diagnosis is intrinsically linked to the quality of fault estimated values that it can provide. Therefore, this chapter is mainly interested in the study of the performance of the metaheuristics when solving the optimization problems in Step 3.

In particular, this chapter is interested in the performance analysis of each one of the metaheuristics, which are used for the diagnosis in the benchmark problems. The study is based on three basic desirable characteristics of diagnosis: robustness, sensitivity, and computational cost.

With that purpose, the chosen experiments consider different faulty situations:

- with a single fault: only a fault can affect the system.
- with multiple faults: more than one fault affect at the same time the system.
- with incipient faults: faults, whose effects in the output are within the rank of the noise affecting the output. In other words, faults of small magnitude.
- without faults.

The measurements are corrupted with different noise levels. The robustness analysis is based on experiments with situations in which the noise is high (up to 8%). The system can be affected by single or multiple faults. The use of high noise levels has the objective of simulating in the output the effect of external disturbances that can eventually affect the system.

The sensitivity analysis is based on experiments with incipient faults and a highly noisy environment (up to 8%).

It is also tested how some parameters of the applied metaheuristics influence the quality of the diagnosis.

Some elements of descriptive statistics, the Sign Test and Wilcoxon's test, are applied in order to analyze the results [30, 45]. The results are presented in different tables and graphs, with the aim of showing different ways of presenting and interpreting the results.

The Sign Test is an easy way to compare the overall performance of two algorithms [30]. For this test, it is analyzed the number of cases that an algorithm is the overall winner. The null hypothesis H_0 considers that both algorithms are equivalent. The number of wins is distributed according to a binomial distribution [30].

The Wilcoxon's test is a simple and safe nonparametric test for pairwise statistical comparisons [30, 45]. The null hypothesis H_0 considers that given two vectors of data $\mathbf{x}$ and $\mathbf{y}$, the vector $\mathbf{d} = \mathbf{x} - \mathbf{y}$ can be represented by a continuous symmetric distribution with null median. In this test, the statistical parameter $T = \min \{R^+, R^-\}$ is computed and compared with the value of the Wilcoxon's distribution for a total of Num degrees of freedom (critical value of W), where Num is the number of cases for which the performance of the algorithms is compared [30]. The ranks R^+ and R^- are computed as follows:

$$R^+ = \sum_{d_i > 0} rank(d_i) + \frac{1}{2} \sum_{d_i = 0} rank(d_i) \tag{4.1}$$

$$R^- = \sum_{d_i < 0} rank(d_i) + \frac{1}{2} \sum_{d_i = 0} rank(d_i) \tag{4.2}$$

where d_i are the elements of the vector $\mathbf{d}$ which provide the difference between the performance of both algorithms for the ith case. These differences should be in ascending order. The assigned position of the component d_i in the new ascending order is taken as $rank(d_i)$.

As stochastic algorithms are of intensive search, with high computational costs, it is necessary to establish the stopping criteria, when the metaheuristics are used. The three stopping criteria used in this book are:

- Criterion 1: Maximum number of iterations allowed, $MaxIter$. This is related to the maximum computational cost which is permitted.
- Criterion 2: Maximum number of iterations for which the best value of the objective function remains constant, Itr_{cte}.
- Criterion 3: Value of the objective function $f(\hat{\boldsymbol{\theta}}_f)$, which is compared with a certain value f_{stop}. The value of f_{stop} is taken as $0.01 y_{max}^2$, being y_{max} the maximum value of all the measured output of the system during the considered sample time. Next it is presented under which reasoning this value was computed.

As described in Chap. 2, the value of the objective function cannot be zero due to a number of reasons, such as model uncertainties, noise in the measurements, and other disturbances. When some prior information about the disturbances is known, for example the level of noise affecting the system, it can be incorporated in the stopping criterion, leading a more realistic value for f_{stop}. For the test cases under investigation in this book, it is considered a small value, $f_{stop} = 0.01 y_{max}^2$, where y_{max} is the maximum value of all the measured output of the system during the considered sample time. A small value for f_{stop} allows to analyze the performance of the metaheuristics when estimating the faults without previous knowledge on the noise level. This value will be used in all the benchmark problems considered. It was computed considering that the number of sampling times $I = 100$, the relative error ϵ between the measured output value and the computed output value is $\epsilon \leq 0.01$ (it means, less or equal than 1%).

Being $y_t^i(\boldsymbol{\theta}_f)$ and $\hat{y}_t^i(\hat{\boldsymbol{\theta}}_f)$, with $i = 1, 2, \ldots p$, the components of the output measured vector $\boldsymbol{Y}_t(\boldsymbol{\theta}_f) \in \mathbb{R}^p$ and the computed output vector $\hat{\boldsymbol{Y}}_t(\hat{\boldsymbol{\theta}}_f)$, respectively, and at the instant of time t, with $t = 1, 2, \ldots, I$. Then for every ith component it is obtained:

$$\frac{y_t^i(\boldsymbol{\theta}_f) - \hat{y}_t^i(\hat{\boldsymbol{\theta}}_f)}{y_t^i(\boldsymbol{\theta}_f)} \leq \epsilon \tag{4.3}$$

Therefore,

$$\left(\frac{y_t^i(\boldsymbol{\theta}_f) - \hat{y}_t^i(\hat{\boldsymbol{\theta}}_f)}{y_t^i(\boldsymbol{\theta}_f)} \right)^2 \leq \epsilon^2 \tag{4.4}$$

and

$$\left(y_t^i(\boldsymbol{\theta}_f) - \hat{y}_t^i(\hat{\boldsymbol{\theta}}_f) \right)^2 \leq \epsilon^2 (y_t^i(\boldsymbol{\theta}_f))^2 \tag{4.5}$$

Considering the maximum absolute value within all the measured values for this component during the considered sample time $y_{max\,i}$, then:

$$\left(y_t^i(\boldsymbol{\theta}_f) - \hat{y}_t^i(\hat{\boldsymbol{\theta}}_f) \right)^2 \leq \epsilon^2 y_{max}^2 \tag{4.6}$$

and

$$\sum_{t=1}^{I} \left[y_t^i(\boldsymbol{\theta}_f) - \hat{y}_t^i(\hat{\boldsymbol{\theta}}_f) \right]^2 \leq I \epsilon^2 y_{max}^2 \tag{4.7}$$

Considering the objective function of the optimization problem for the Fault Diagnosis Inverse Problem, see Eq. (2.2), it is obtained:

$$f(\hat{\boldsymbol{\theta}}_f) = \left\| \sum_{t=1}^{I} \left[\boldsymbol{Y}_t(\boldsymbol{\theta}_f) - \hat{\boldsymbol{Y}}_t(\hat{\boldsymbol{\theta}}_f) \right]^2 \right\|_{\infty} \leq I \epsilon^2 y_{max}^2 \tag{4.8}$$

For $I = 100$ and a small relative error $\epsilon = 0.01$, it is obtained: $f_{stop} = 0.01 y_{max}^2$. The computation of f_{umbral} in Step 2 can be made following the same criteria, see Sect. 4.3.

As a remark, Criterion 1 was included explicitly in the description of all algorithms presented in Chap. 3. All implementations, whose results are presented in this book, were made in Matlab®. The computer used during all the experiments has a processor Inter(R) Core(TM) i7-4500U CPU @ 1.80 GHz 2.40 GHz and a RAM of 8 GB. The codes for the metaheuristics DEwPC and PSO-M are presented in Appendices A and B, respectively.

4.2   Experiments with the DC Motor

This benchmark problem satisfies hypothesis H1–H3 of FD-IPM, as shown in Chap. 2. Therefore, the faults can be diagnosed with FD-IPM. The algorithms PSO, ACO, and PSO-M were applied in Step 3, i.e. the three metaheuristics were applied to solve the optimization problem that results from the application of Step 1 of FD-IPM to the DC Motor described in Chap. 2, see Eq. (2.36). The experiments for this benchmark problem were divided into three parts:

- *General analysis*
  Designed for the general analysis of the diagnosis based on PSO and ACO. This part includes multiple faults situations (Cases 1–4 from Table 4.2). The output of the system is corrupted up to a 2% noise level.
- *Robustness analysis*
  Same situations as in the *General Analysis*, but with a noise level up to 8%.
- *Sensitivity analysis*
  The cases include simple and incipient faults, Cases 6–8 from Table 4.2, as well as multiple and incipient faults, see Case 5 from Table 4.2. The measurements are corrupted with a noise level up to 8%.

Table 4.2 Faulty situations considered for the experiments with the DC Motor benchmark system

Case	f_u	f_y	f_p
1	0.87	−0.12	0.53
2	−0.27	0.96	0
3	0.63	0	0.29
4	0	0.47	0.86
5	−0.08	0.09	0.2
6	0.15	0	0
7	0	−0.1	0
8	0	0	0.12

In order to analyze the influence of some of the PSO and ACO algorithms parameters in the diagnosis, different values for such parameters were considered. In order to obtain statistically valid conclusions, each variant of both algorithms was run 30 times, for each faulty situation under study. A descriptive statistics parameter was computed: mean value of the fault estimates. This statistics provides a measure of central tendency of the estimations for each fault. The notations used in the tables and figures presented in this chapter were: $\bar{f}(\hat{\boldsymbol{\theta}}_f)$ for the mean value of the objective function; $\overline{Eval}$ for the mean of the objective function number of evaluations performed until the minimum value of the objective function is achieved; $\bar{f}_u$, $\bar{f}_y$, and $\bar{f}_p$ for the mean values of the estimates for the faults f_u, f_y and f_p, respectively.

In these algorithms, the evaluation of the objective function represents the highest cost in the computational aspect. Every evaluation of the objective function implies the solution of the direct problem, see model described by Eq. (2.19). Therefore, the metrics related to the computational cost in this study is based on $\overline{Eval}$.

The best set of parameters for PSO and ACO, respectively, were selected based on the results from the experiments performed. For that, Sign Test was applied to the results.

The conclusion about the comparison between the best variant of PSO and the best variant of ACO, and between PSO and PSO-M, was based on the Wilcoxon's test.

4.2.1 Implementations

- Implementation of PSO

    Two variants of PSO algorithm were implemented: PSO Canonical (PSO_B), and PSO with inertial weight (PSO_I). The fundamental difference among them is the way the velocity of each particle is updated, see Sect. 3.5, which influences the trend of the search.

    For PSO with inertial weight, it was considered different recommended values for the parameters c_1 and c_2 [23, 65, 70]. These parameters also determine the balance diversification-intensification of the search. Therefore, it can be analyzed the influence of the trend of the search in the quality of the diagnosis. In Table 4.3 are presented the different values considered for the parameters, and the name adopted for each variant of the PSO algorithm. In Table 4.3 the number of the current iteration is denoted by $Iter$, the maximum number of birds is Z, and the maximum number of iterations to be performed by the algorithms is $MaxIter$. The implementation was based on the algorithm shown in Algorithm 6.

- Implementation of ACO

    For ACO, the parameters q_0 and K were selected by analyzing their influence in the quality of the diagnosis. Parameter q_0 determines the level of randomness in the selection of the discrete value for each of the known variables, i.e. in the

Table 4.3 Parameter values used for the PSO algorithm in the experiments with the DC Motor benchmark problem

Notation	ω	c_1	c_2	Z	ω_{max}	ω_{min}
PSO_B	1	2	2	30	–	–
PSO_{I1}	$\omega_{max} - \frac{\omega_{max}-\omega_{min}}{MaxIter}Iter$	2	2	30	0.4	0.9
PSO_{I2}	$\omega_{max} - \frac{\omega_{max}-\omega_{min}}{MaxIter}Iter$	3	1	30	0.4	0.9
PSO_{I3}	$\omega_{max} - \frac{\omega_{max}-\omega_{min}}{MaxIter}Iter$	3.5	0.5	30	0.4	0.9

Table 4.4 Parameter values used for the ACO algorithm in the experiments with the DC Motor benchmark problem

Notation	K	q_0	Z	C_{inc}	C_{evap}
$ACO_{1.1}$	63	0.15	30	0.30	0.10
$ACO_{1.2}$	63	0.55	30	0.30	0.10
$ACO_{1.3}$	63	0.85	30	0.30	0.10
$ACO_{2.1}$	127	0.15	30	0.30	0.10
$ACO_{2.2}$	127	0.55	30	0.30	0.10
$ACO_{2.3}$	127	0.85	30	0.30	0.10

process of ant generation [111]. Therefore, its value influences the trend of the search. For this analysis, the selected values for q_0 were:

– $q_0 = 0.15$: this value determines a greater diversification of the search.
– $q_0 = 0.55$: this value determines a balance between diversification and intensification.
– $q_0 = 0.85$: this value determines a greater intensification of the search.

For parameter K that determines the size of the search space, by defining the number of discrete values used for each unknown, two values were used: 63 and 127.

In Table 4.4 are presented the different values considered for the parameters, and the name adopted for each variant of the ACO algorithm. The value of Z represents the number of ants, C_{inc} and C_{evap} are the incremental factor and the evaporation factor, respectively. The implementation was based on the algorithm shown in Algorithm 5.

• Implementation of PSO-M

The implementation of PSO-M is based on the algorithm shown in Algorithm 8. The values of the PSO-M parameters used during its implementation were selected based on the recommendations shown in Table 3.10. In Table 4.5 are presented the values considered for the PSO-M parameters in its application to the DC Motor benchmark problem.

• Stopping Criteria

The stopping criteria used for PSO and ACO algorithms were:

– **Criterion 1:** Maximum number of iterations: $MaxIter = 100$.
– **Criterion 2:** Maximum number of iterations for which the best value of the objective function remains constant: $Itr_{cte} = 10$.

Table 4.5 Parameter values used for the PSO-M algorithm in the experiments with the DC Motor benchmark problem

First stage

Z_1	c_1	c_2	ω	$MaxIter_1$	K	q_0	C_{inc}	C_{evap}
45	2.05	2.05	0.729	10	63	0.55	0.30	0.10

Second stage

Z_2	c_1	c_2	ω	$MaxIter_2$				
15	2.05	2.05	0.729	90				

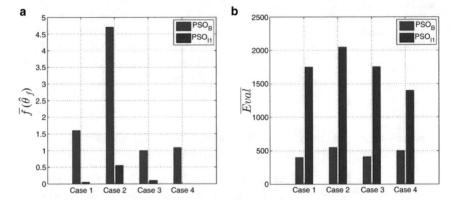

Fig. 4.1 Performance comparison for PSO_B and PSO_{I1} when diagnosing faults in Cases 1–4 from Table 4.2, using data with up to 2% noise level

- **Criterion 3:** Value of the objective function $f(\hat{\boldsymbol{\theta}}_f) \leq 0.01 y_{max}^2$, being y_{max} the maximum value of all the measured output of the system during the considered sampling time.

For the first stage of PSO-M, it was only considered stopping criterion number 1, with $MaxIter = 10$. In the second stage of PSO-M, criteria 1 and 3 were considered, with $MaxIter_2 = 90$..

In the results presented next, at the end of the experiments with the algorithms, it is also presented a comparison with other FDI model based methods.

4.2.2 Results of the Diagnosis with Particle Swarm Optimization

4.2.2.1 General Analysis

In Fig. 4.1 it is shown that both variants PSO_B and PSO_{I1} (see Table 4.3) detect the faults for the Cases 1–4 (see Table 4.2), in the sense that their estimations are different from zero. Based on the average value of the objective function $\bar{f}(\hat{\boldsymbol{\theta}}_f)$, see

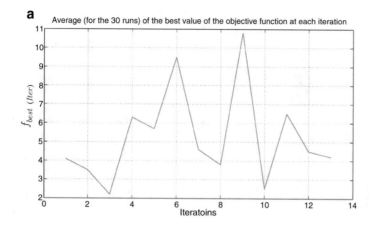

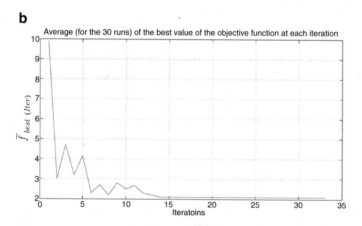

Fig. 4.2 Average (for the 30 runs) of the best value of the objective function at each iteration $\bar{f}_{\text{best}(Iter)}$, obtained with (**a**) PSO$_B$ and (**b**) PSO$_{I1}$ for the Case 3 from Table 4.2, using data with up to 2% noise level

Fig. 4.1a, it can be concluded that PSO$_{I1}$ is more precise in the estimations of the faults than PSO$_B$. Based on the average number of function evaluations $\overline{Eval}$, see Fig. 4.1b, it can be concluded that PSO$_{I1}$ has a higher computational cost.

With the objective to illustrate better the previous results just presented, Case 3 from Table 4.2 will be analyzed in detail. In Fig. 4.2 it is shown the average (for the 30 runs) of best value of the objective function at each iteration $\bar{f}_{\text{best}(Iter)}$. From the results shown in Fig. 4.2b it is possible to observe the effect of reducing the value of the inertial weight ω as a function of the number of iterations with the algorithm PSO$_{I1}$. This indicates a higher intensification around the better solutions. For this reason, PSO$_{I1}$ executes more evaluations of the objective function than PSO$_B$. However, PSO$_{I1}$ gives better fault estimations. Note in Fig. 4.2b how the

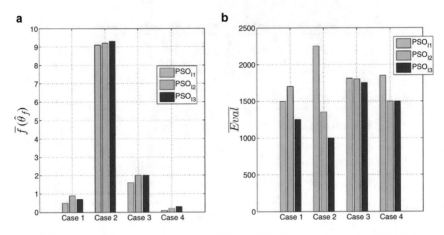

Fig. 4.3 Performance comparison for PSO_{I1}, PSO_{I2}, and PSO_{I3} when diagnosing faults in Cases 1–4 from Table 4.2, using data with up to 8% noise level

final value of the objective function is closer to zero. That seems to be related with how the algorithm explores the search space due to the inertial weight, which allows to modify the trend of the search.

Taking into account the results just presented, in the next section will be only considered the variants of PSO with different values for parameters c_1 and c_2, see Table 4.3.

4.2.2.2   Robustness Analysis

The variants PSO_{I1}, PSO_{I2}, and PSO_{I3}, from Table 4.3, were applied to diagnosing faults in the DC Motor benchmark problem. In Fig. 4.3a it is shown that there are no big differences in the estimates that each algorithm provides. Therefore, no big differences in the robustness of the diagnosis are obtained. However, in Fig. 4.3b there are clear differences with respect to the computational cost involved. It can be observed that PSO_{I3} required, in most of the experiments, the lowest number of objective function evaluations. It can be concluded that parameters c_1 and c_2 do not have a great influence in the robustness, but they have a significant influence on the computational cost.

4.2.2.3   Sensitivity Analysis

In Fig. 4.4a it is shown that the worst result in the estimation of the faults for the Cases 5–8, see Table 4.2, is obtained with PSO_{I3}, while PSO_{I1} and PSO_{I2} provide similar results. In Fig. 4.4b it is shown that there is no a clear trend regarding the computational cost. Each PSO variant performs better in one case, but requires more function evaluations in the others.

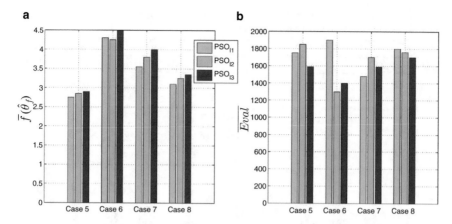

Fig. 4.4 Performance comparison for PSO_{I1}, PSO_{I2}, and PSO_{I3} when diagnosing faults in Cases 5–8 from Table 4.2, using data with up to 8% noise level

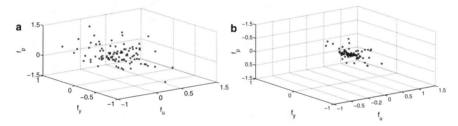

Fig. 4.5 Comparison between the search of (**a**) PSO_{I1} and (**b**) PSO_{I3} for Case 5 in Table 4.2

In order to analyze the effect of diversification on the sensitivity, Fig. 4.5 shows a comparison between the search with PSO_{I1} and PSO_{I3} along the iterative procedure for the Case 5 in Table 4.2. As observed in Fig. 4.5a, PSO_{I1} performs a greater diversification than PSO_{I3}. Therefore, it can be concluded from the results presented in Figs. 4.4a and 4.5 that a greater diversification in the search could be important for obtaining a sensitive diagnosis.

Taking into account all these results, it is possible to conclude that the best variants for obtaining a sensitive diagnosis are PSO_{I1} and PSO_{I2}. Based on the Sign Test (see Sect. 4.1) it is chosen PSO_{I1} as the best variant with a significance level $\alpha = 0.05$. For this test, were considered Cases 1–8 from Table 4.2 using data up to 8% noise level. The results of the Sign Test are presented in Table 4.6. The number of cases used for the application of the test is denoted by *Num* and *lost* denotes the number of cases for which the algorithm was the worst.

Table 4.6 Results of Sign Test: PSO_{I1} versus PSO_{I2}, when diagnosing fault Cases 1–8 from Table 4.2, using data with up to 8% noise level

Comparison	Criterion	Wins	Lost	Num	Critical value	α
PSO_{I1} vs PSO_{I2}	$\bar{f}(\hat{\theta}_f)$	7	1	8	7	0.05

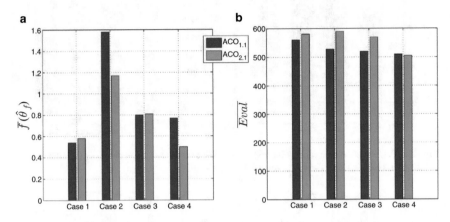

Fig. 4.6 Performance comparison for $ACO_{1.1}$ and $ACO_{2.1}$, taking as criterium (**a**) mean of the value of the objective function ($\bar{f}(\hat{\theta}_f)$) and (**b**) mean of the number of function evaluations ($\overline{Eval}$), in both cases diagnosing faults in Cases 1–4 from Table 4.2, using data with up to 2% noise level

4.2.3 Results of the Diagnosis with Ant Colony Optimization

4.2.3.1 General Analysis

The variants $ACO_{1.1}$ and $ACO_{2.1}$, see Table 4.4, are considered in this section. The difference between these two variants is the value of the parameter K. In Fig. 4.6 are shown the results of the faults diagnosis for the test Cases 1–2 in Table 4.2. Both variants were able to detect the faults. Variant $ACO_{1.1}$ shows the best performance for Case 1: better value of the objective function and lower computational cost than $ACO_{2.1}$. For Case 2, $ACO_{2.1}$ obtains a better value of the objective function, but with a higher computational cost than $ACO_{1.1}$. For Case 3 the values of the objective functions obtained by both algorithms are similar, but the computational cost is higher for $ACO_{2.1}$. For Case 4 $ACO_{2.1}$ is better than $ACO_{1.1}$ with respect to both criteria: lower value for the objective function, and smaller average number of function evaluations.

In Fig. 4.7 it is shown a comparison between $ACO_{1.1}$ and $ACO_{2.1}$, based on the evolution of the best value of the objective function obtained at each iteration for Case 3 from Table 4.2, using data with up to 2% noise level. The greater discretization of the search space of $ACO_{2.1}$, with a low value for parameter q_0, produces smaller variations in the value of the objective function along the each iteration of the algorithm than $ACO_{1.1}$. Therefore, $ACO_{2.1}$ executes a higher number of function evaluations, which means a higher computational cost than $ACO_{1.1}$, for similar fault estimations.

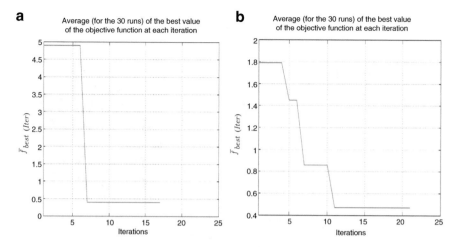

Fig. 4.7 Comparison between the average of the best values of the objective function $\bar{f}_{\text{best}(Iter)}$ at each iteration obtained with (**a**) $ACO_{1.1}$ and (**b**) $ACO_{2.1}$, when diagnosing faults in Case 3 from Table 4.2, using data with up to 2% noise level

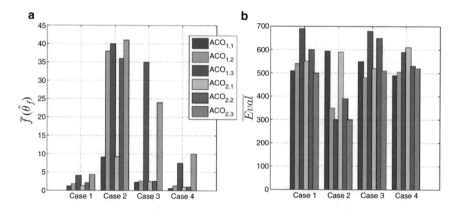

Fig. 4.8 Performance comparison obtained for the six variants of ACO listed in Table 4.4, taking as criteria (**a**) mean of the value of the objective function $(\bar{f}(\hat{\boldsymbol{\theta}}_f))$ and (**b**) mean of the number of function evaluations $(\overline{Eval})$, in both cases diagnosing faults in Cases 1–4 from Table 4.2, using data with up to 8% noise level

4.2.3.2   Robustness Analysis

In Fig. 4.8 it is shown a performance comparison for the six variants of ACO listed in Table 4.4, when applied to Cases 1–4 in Table 4.2. The results fluctuated within a wide range for some of these variants. Variants $ACO_{1.3}$ and $ACO_{2.3}$, which correspond to the algorithms with the higher intensification due to the higher value $q_0 = 0.85$, give the worst estimations. The estimates obtained with $ACO_{1.1}$ and $ACO_{2.1}$, which correspond to the algorithms with the higher diversification due to

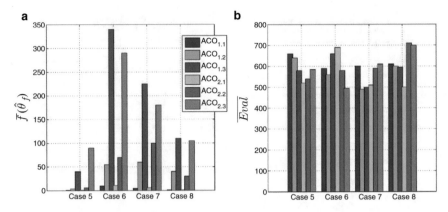

Fig. 4.9 Performance comparison obtained for the six variants of ACO listed in Table 4.4, taking as criteria (**a**) mean of the value of the objective function ($\bar{f}(\hat{\boldsymbol{\theta}}_f)$) and (**b**) mean of the number of function evaluations ($\overline{Eval}$), in both cases, diagnosing Cases 5–8 from Table 4.2, using data with up to 8% noise level

the lower value $q_0 = 0.15$, are more accurate than those obtained with $ACO_{1.2}$ and $ACO_{2.2}$, respectively, as can be observed in Fig. 4.8a. It can be concluded that the higher diversification obtained with the lower value $q_0 = 0.15$ has a good influence on the robustness of the diagnosis. Regarding the computational cost, there is no a clear trend.

4.2.3.3 Sensitivity Analysis

In Fig. 4.9 it is shown a performance comparison for the six variants of ACO under study, when applied to Cases 5–8 in Table 4.2. As in the robustness analysis performed in the previous section, $ACO_{1.1}$ and $ACO_{2.1}$ give the most accurate estimates. Thus, it can be concluded that the higher diversification obtained with the lower parameter value $q_0 = 0.15$ has also a good influence on the diagnosis sensitivity. Concerning the computational cost, it is observed in Fig. 4.9b that $ACO_{2.1}$ is the best in two out of the four cases under study, i.e. Cases 5 and 8.

In order to analyze the effect of diversification on the sensitivity, Fig. 4.10 shows a comparison of the search for the six variants of ACO, along the iterative procedure, when diagnosing faults in Case 5 from Table 4.2, using data with up to 8% noise level. The figures show that the variants $ACO_{1.1}$ and $ACO_{2.1}$ perform a greater diversification than the other. It can be concluded that a greater diversification in the search could be important for obtaining a sensitive diagnosis.

Taking into consideration the results of the previous analysis, it is possible to conclude that the best variants of ACO are $ACO_{1.1}$ and $ACO_{2.1}$. The application of the Sign Test, see Sect. 4.1, confirmed that the best variant is $ACO_{1.1}$, with a significance level $\alpha = 0.05$. Table 4.7 shows for the results.

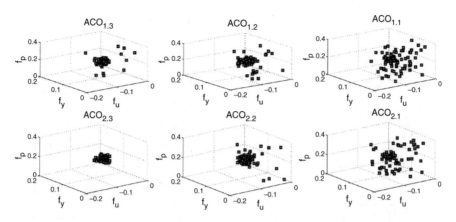

Fig. 4.10 Comparison among the search for the six variants of ACO listed in Table 4.4 when diagnosing faults in Case 5 from Table 4.2, using data with up to 8% noise level

Table 4.7 Results of Sign Test: $ACO_{1.1}$ versus $ACO_{2.1}$, when diagnosing faults in Cases 1–8 from Table 4.2, using data with up to 8% noise level

Comparison	Criterion	Wins	Lost	Num	Critical value	α
$ACO_{1.1}$ vs $ACO_{2.1}$	$\bar{f}(\hat{\theta}_f)$	7	1	8	7	0.05

Based on the previous study, PSO_{I1} and $ACO_{1.1}$ were chosen as the best variants of PSO and ACO, respectively. The numerical values obtained for the faults with both algorithms, during the experiments, are presented in Table 4.8. In Table 4.8 the bold values are used to emphasize the best numerical value obtained between PSO_{I1} and $ACO_{1.1}$ for each criterium represented in this table at every faulty case under study.

The results presented in Table 4.8 show that PSO_{I1} gives better fault estimations than $ACO_{1.1}$ in six out of the eight cases considered. In Fig. 4.11 it is shown that the number of objective function evaluations needed by $ACO_{1.1}$ is smaller than what is needed by PSO_{I1}, while the value of the objective function is always smaller in the diagnosis with PSO_{I1}.

Table 4.9 shows the results of the application of the Wilcoxon's test, see Sect. 4.1. For the first line R^+ represents the sum of the ranks for which PSO_{I1} outperformed $ACO_{1.1}$, taking as the comparison metrics the average value of the objective function. For the second line R^+ represents the sum of the ranks for which $ACO_{1.1}$ outperformed PSO_{I1}, taking as the comparison metrics the average of the number of objective function evaluations.

From the results shown in Table 4.9, it can be concluded that PSO_{I1} shows a significant improvement over the estimations obtained by $ACO_{1.1}$, with a significance level $\alpha = 0.01$. On the other hand, $ACO_{1.1}$ permits to obtain acceptable estimates with a computational cost significantly lower, when compared to PSO_{I1}, with the same significance level $\alpha = 0.01$. This issue is important for online diagnosis.

Table 4.8 Results for the fault diagnosis with $ACO_{1.1}$ and PSO_{I1}, using data with up to 8% noise level

Case	Variant	$\bar{f}_u$	$\bar{f}_y$	$\bar{f}_p$	$\bar{f}(\hat{\boldsymbol{\theta}}_f)$	$\overline{Eval}$
$(f_u = 0.87)$ $(f_y = -0.12)$ $(f_p = 0.53)$						
1	$ACO_{1.1}$	0.8508	−0.0794	0.5429	1.4702	**510**
	PSO_{I1}	0.8778	−0.0985	0.5402	**0.7018**	1491
$(f_u = -0.27)$ $(f_y = 0.96)$ $(f_p = 0)$						
2	$ACO_{1.1}$	−0.2349	0.7556	−0.0857	9.5828	**591**
	PSO_{I1}	−0.2496	0.9944	0.01734	**9.1491**	2244
$(f_u = 0.63)$ $(f_y = 0)$ $(f_p = 0.29)$						
3	$ACO_{1.1}$	0.6159	0.0889	0.3270	2.4367	**549**
	PSO_{I1}	0.6428	0.0275	0.3040	**2.1905**	1734
$(f_u = 0)$ $(f_y = 0.47)$ $(f_p = 0.86)$						
4	$ACO_{1.1}$	−0.0063	0.4889	0.8667	1.0940	**483**
	PSO_{I1}	−0.0010	0.4576	0.8545	**0.3261**	1761
$(f_u = -0.08)$ $(f_y = 0.09)$ $(f_p = 0.2)$						
5	$ACO_{1.1}$	−0.1397	0.0444	0.1746	3.3207	**660**
	PSO_{I1}	−0.0775	0.0951	0.2023	**2.8132**	1716
$(f_u = 0.15)$ $(f_y = 0)$ $(f_p = 0)$						
6	$ACO_{1.1}$	0.1714	0.0349	0.0222	4.9194	**576**
	PSO_{I1}	0.1633	0.0175	0.0088	**4.3135**	1941
$(f_u = 0)$ $(f_y = -0.1)$ $(f_p = 0)$						
7	$ACO_{1.1}$	0.0032	−0.2127	−0.0508	4.3432	**606**
	PSO_{I1}	−0.0103	−0.0884	0.0043	**3.7786**	1473
$(f_u = 0)$ $(f_y = 0)$ $(f_p = 0.12)$						
8	$ACO_{1.1}$	−0.0190	0.0063	0.1175	3.6654	**633**
	PSO_{I1}	−0.0028	0.0135	0.1261	**3.1340**	1830

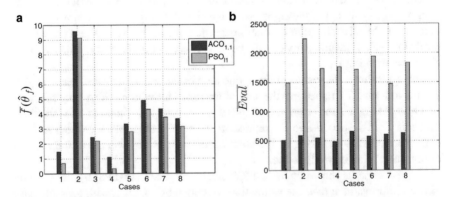

Fig. 4.11 Performance comparison for $ACO_{1.1}$ and PSO_{I1}, taking as criteria (**a**) mean of the value of the objective function ($\bar{f}(\hat{\boldsymbol{\theta}}_f)$) and (**b**) mean of the number of function evaluations ($\overline{Eval}$), in both cases, diagnosing faulty situations from Table 4.2, using data with up to 8% noise level

Table 4.9 Wilcoxon's test results: PSO_{I1} versus $ACO_{1.1}$

Comparison	Criterion	R^+	R^-	$T = \min \{R^+, R^-\}$	Critical value of W	α
PSO_{I1} vs $ACO_{1.1}$	$\bar{f}(\hat{\boldsymbol{\theta}}_f)$	36	0	0	0	0.01
$ACO_{1.1}$ vs PSO_{I1}	$\overline{Eval}$	36	0	0	0	0.01

4.2.4 Diagnosis with Particle Swarm Optimization with Memory (PSO-M)

With the application of the metaheuristic PSO-M, it is expected to be kept the quality of the estimations provided by PSO_{I1} (which means robust and sensitive diagnosis), while reducing the number of objective function evaluations, using the pheromone strategy from ACO in that sense, which can be understood as a learning mechanism (see Sect. 3.7). This approach leads to an improvement of the PSO_{I1} performance for online diagnosis.

PSO-M was applied for diagnosing faults in the same cases that were considered in the previous robustness and sensitivity analysis. The numerical results are presented in Table 4.10. The results obtained with PSO_{I1} and $ACO_{1.1}$ are also presented, in order to allow a direct comparison. The bold values are used to emphasize the best numerical value obtained at each case. Table 4.10 shows that PSO-M obtains more accurate fault estimates in six out of the eight cases. The algorithm PSO-M also reduces the number of function evaluations required by PSO_{I1} in all cases considered.

In order to evaluate the performance of PSO-M, it was applied the Wilcoxon' test for the comparison between PSO_{I1} and PSO-M. In Table 4.11 the results are presented. R^+ represents the sum of the ranks for which PSO-M outperformed PSO_{I1}. That holds for both lines in Table 4.11. The value of the objective function and the number of objective function evaluation were chosen as the comparison metrics.

The results on the first line of Table 4.11 indicate that the null hypothesis cannot be rejected, which means that there are no differences between the fault estimations performed with PSO-M and PSO_{I1}. The results from the second line indicate that there is an over performance of PSO-M with respect to PSO_{I1}, with a significance level $\alpha = 0.01$, which implies less computational cost.

The performance comparison among PSO_{I1}, $ACO_{1.1}$, and PSO-M is also shown in Fig. 4.12. The algorithm PSO-M obtains similar estimates to PSO_{I1}, as the Wilcoxon's test validated, and it also permits to reduce the computational time required. Therefore, PSO-M improves the diagnosis in comparison to PSO and ACO.

Table 4.10 Results for the Fault Diagnosis with $ACO_{1.1}$, PSO_{I1}, and PSO-M, using data with up to 8% noise level

Case	Algorithm	$\bar{f}_u$	$\bar{f}_y$	$\bar{f}_p$	$\bar{f}(\hat{\boldsymbol{\theta}}_f)$	$\overline{Eval}$
$(f_u = 0.87)$ $(f_y = -0.12)$ $(f_p = 0.53)$						
1	$ACO_{1.1}$	0.8508	−0.0794	0.5429	1.4702	510
	PSO_{I1}	0.8778	−0.0985	0.5402	0.7018	1491
	PSO-M	0.8635	−0.1496	0.5526	**0.5966**	1037
$(f_u = -0.27)$ $(f_y = 0.96)$ $(f_p = 0)$						
2	$ACO_{1.1}$	−0.2349	0.7556	−0.0857	9.5828	591
	PSO_{I1}	−0.2496	0.9944	0.01734	9.1491	2244
	PSO-M	−0.2995	0.9587	−0.0015	**9.0020**	1083
$(f_u = 0.63)$ $(f_y = 0)$ $(f_p = 0.29)$						
3	$ACO_{1.1}$	0.6159	0.0889	0.3270	2.4367	549
	PSO_{I1}	0.6428	0.0275	0.3040	2.1905	1734
	PSO-M	0.6412	0.0859	0.3293	**2.1881**	991
$(f_u = 0)$ $(f_y = 0.47)$ $(f_p = 0.86)$						
4	$ACO_{1.1}$	−0.0063	0.4889	0.8667	1.0940	483
	PSO_{I1}	−0.0010	0.4576	0.8545	**0.3261**	1761
	PSO-M	0.0087	0.4614	0.8570	0.3548	947
$(f_u = -0.08)$ $(f_y = 0.09)$ $(f_p = 0.2)$						
5	$ACO_{1.1}$	−0.1397	0.0444	0.1746	3.3207	660
	PSO_{I1}	−0.0775	0.0951	0.2023	**2.8132**	1716
	PSO-M	−0.0720	0.1126	0.2113	2.8339	1051
$(f_u = 0.15)$ $(f_y = 0)$ $(f_p = 0)$						
6	$ACO_{1.1}$	0.1714	0.0349	0.0222	4.9194	576
	PSO_{I1}	0.1633	0.0175	0.0088	4.3135	1941
	PSO-M	0.1483	0.0247	0.0109	**4.3073**	963
$(f_u = 0)$ $(f_y = -0.1)$ $(f_p = 0)$						
7	$ACO_{1.1}$	0.0032	−0.2127	−0.0508	4.3432	606
	PSO_{I1}	−0.0103	−0.0884	0.0043	3.7786	1473
	PSO-M	−0.0039	−0.0547	0.0194	**3.6969**	1077
$(f_u = 0)$ $(f_y = 0)$ $(f_p = 0.12)$						
8	$ACO_{1.1}$	−0.0190	0.0063	0.1175	3.6654	633
	PSO_{I1}	−0.0028	0.0135	0.1261	3.1399	1830
	PSO-M	0.0118	0.0315	0.1358	**3.1353**	1109

Table 4.11 Results for the Wilcoxon's test: PSO-M versus PSO_{I1}

Comparison	Criterion	R^+	R^-	$T = \min\{R^+, R^-\}$	Critical value of W	α
PSO-M vs PSO_{I1}	$\bar{f}(\hat{\boldsymbol{\theta}}_f)$	27	9	9	6	0.1
PSO-M vs PSO_{I1}	$\overline{Eval}$	36	0	0	0	0.01

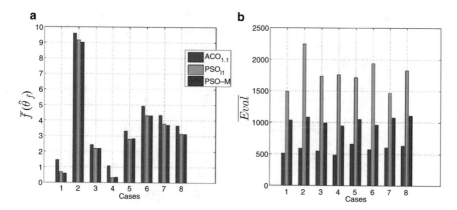

Fig. 4.12 Performance comparison for $ACO_{1.1}$, PSO_{l1}, and PSO-M, taking as criteria (**a**) mean of the value of the objective function ($\bar{f}(\hat{\theta}_f)$) and (**b**) mean of the number of function evaluations ($\overline{Eval}$), in both cases, diagnosing faults in Cases 1–8 in Table 4.2, and using data with up to 8% noise level

4.2.5   Comparison of the Fault Diagnosis: Inverse Problem Methodology with Other FDI Model Based Methods

A comparison with other FDI model based methods is presented in this section. As described in Sect. 1.1, the main model based approaches for FDI are Parity Space and Diagnostic Observers. Therefore, Parity Space and Diagnostic Observers are used next, for comparison purposes with FD-IPM.

Both approaches generate residuals, which are used to construct appropriate decision functions.

The idea is to generate a structured set of residuals, such that at least one measured quantity has no impact on a specific residue. In case of a faulty measurement, the decoupled residue remains small, while the others increase their value. This feature helps to locate the fault [42, 55].

Both approaches are formulated based on the state representation, but in the linear case it can be connected with the transfer function models [31].

- Parity Space approach:

For the intended comparison, it is considered the original form of Parity Space, which is based on the assumption of a state space model of a linear discrete time system [25] in the form:

$$\begin{aligned}
x\,(k+1) &= Ax(k) + Bu(k) + E_f\theta_f(k) \\
y\,(k) &= Cx(k) + F_f\theta_f(k)
\end{aligned} \tag{4.9}$$

where $x \in \mathbb{R}^n$ is the vector of state variables; $u \in \mathbb{R}^m$ and $y \in \mathbb{R}^p$ the measurable input and output signals, respectively. The matrices A, B, C, E_f, and F_f are known, and with appropriate dimensions.

It is further assumed that (C, A) is observable and $rank(C) = p$. A continuous time parity space can be generated in a similar way to the design of the discrete one [54].

Using the notations:

$$y_s(k) = [y(k-s)\ y(k-s+1)\ldots y(k)]^T ; \tag{4.10}$$

$$u_s(k) = [u(k-s)\ u(k-s+1)\ldots u(k)]^T ; \tag{4.11}$$

$$\theta_{f,s}(k) = [\theta_f(k-s)\ \theta_f(k-s+1)\ldots\theta_f(k)]^T ; \tag{4.12}$$

$$H_{u,s} = \begin{bmatrix} D & 0 & \cdots & 0 \\ CB & D & \ddots & \vdots \\ \vdots & \ddots & \ddots & 0 \\ CA^{s-1}B & \cdots & CB & D \end{bmatrix} \tag{4.13}$$

and

$$H_{f,s} = \begin{bmatrix} F_f & 0 & \cdots & 0 \\ CE_f & F_f & \ddots & \vdots \\ \vdots & \ddots & \ddots & 0 \\ CA^{s-1}E_f & \cdots & CE_f & F_f \end{bmatrix} \tag{4.14}$$

The matrix $H_{o,s}$ is called the extended observability matrix of the system. It is defined as [31]:

$$H_{o,s} = [C\ CA \cdots CA^s]^T \tag{4.15}$$

The parity relation is obtained by:

$$y_s(k) = H_{o,s}x(s-k) + H_{u,s}u_s(k) + H_{f,s}\theta_{f,s}(k) \tag{4.16}$$

If $s \geq n$ exists at least one vector $v_s \in \mathbb{R}^{p \times (s+1)}$ (parity vector) with $v_s \neq 0$ such that $v_s H_{o,s} = 0$. The residual generator based on parity relation, in the absence of faults, is constructed by:

$$r_s(k) = v_s H_{o,s}x(k-s) = v_s(y_s(k) - H_{u,s}u_s(k)) = 0 \tag{4.17}$$

In the presence of faults, the residual becomes:

$$r_s(k) = v_s H_{o,s}x(k-s) + v_s H_{f,s}\theta_{f,s}(k), \quad v_s : v_s H_{o,s} = 0 \tag{4.18}$$

$$r_s(k) = v_s H_{f,s}\theta_{f,s}(k) \neq 0 \tag{4.19}$$

- Diagnostic Observers:

    The basic idea of the observer approach is to reconstruct the state of the system from the measurements of its input and output. The estimation error can be used as the residual for the fault diagnosis [31, 42].

    In the case of a linear time invariant process with state equations

$$\dot{x}(t) = Ax(t) + Bu(t) + E_f\theta_f(t)$$
$$y(t) = Cx(t) + F_f\theta_f(t) \tag{4.20}$$

the estimated state $\hat{x}$ obtained at the output of a full-order observer is governed by the following equations:

$$\dot{\hat{x}} = (A - HC)\hat{x} + Bu + Hy$$
$$\hat{y} = C\hat{x} \tag{4.21}$$

where H denotes the feedback gain matrix that has to be chosen properly in order to achieve a desired performance of the observer.

The relations for the state estimation error, $\xi = x - \hat{x}$, and the output estimation error, $e = y - \hat{y}$, become:

$$\dot{\xi} = (A - HC)\xi + E_f\theta_f + HF_f\theta_f$$
$$e = C\xi + F_f\theta_f \tag{4.22}$$

where e is used as the residual, r, for the purpose of detection and isolation of faults.

- Connections between Parity Space and Diagnostic Observers:

    It has been already demonstrated that exists a one-to-one mapping between the design parameters of diagnostic observers and parity relation based residual generators [31]. It has also been shown that for a given residual generator based on parity relation there exists a set of corresponding observer-based residual generators. It is also known how to calculate the respective parity vector when an observer-based residual generator is provided, and vice-versa [31].

    The system design based on parity space is characterized by its simple mathematical handling. It only deals with matrix and vector valued operations. In the case of observers, the design is more complex. Due to the connection between the parity space approach and observer-based approach, a strategy called parity space design-observer-based implementation was developed [31]. This strategy makes use of the computational advantage of the parity space approach for the system design (selection of a parity vector or matrix), and then realizes the solution in the observer form to ensure a numerically stable and less consuming online computation.

4.2.5.1   Results of the Comparison Between Parity Space, Diagnostic Observers and Fault Diagnosis: Inverse Problem Methodology

In the DC Motor benchmark problem, as shown in Sect. 2.4, one seeks to diagnose three faults with only one output (one sensor). Parity Space and Diagnostic Observers do not permit to diagnose (detect and isolate) additive faults when their number is larger than the number of the sensors [31]. Therefore, the system cannot be diagnosed with Parity Space or Diagnostic Observers, only the fault detection is possible. That is the first advantage of FD-IPM: the three faults can be diagnosed using only one sensor.

Nonetheless, considering that only one fault can affect the system, then it is possible to make a comparison among FD-IPM, Parity Space and Diagnostic Observers. The actuator fault f_u is selected. In order to determine the parity vector and to design the observer, the model of the DC Motor is transformed into its state space representation. This permits to compute the elements for both approaches in an easier way. In the case of FD-IPM, it can be applied independently on the type of model used to describe the system and no further elements need to be designed or computed. The algorithm PSO-M was used for the fault estimations.

The parity vector $v_s = [0.1057 \; -0.4499 \; 0.7162 \; -0.5056 \; 0.1336]^T$ and matrix $H = [-194.1163 \; -1.8394 \; 0.021 \; -1.3084]^t$ were computed.

- Robustness analysis:

  In Fig. 4.13 are shown the residuals resulting from the application of Parity Space and Diagnostic Observers when no faults are affecting the system, and the output of the system is not affected by noise. Under these circumstances, the residuals are equal to zero.

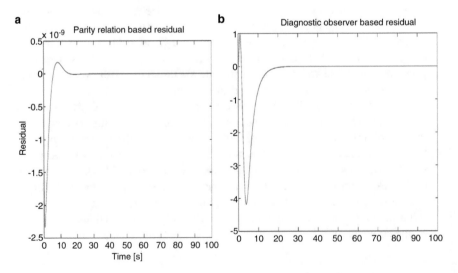

Fig. 4.13 Residuals obtained with (**a**) Parity Space and (**b**) Diagnostic Observers, under the conditions of no faults affecting the system ($f_u = 0$), and no noise affecting the output

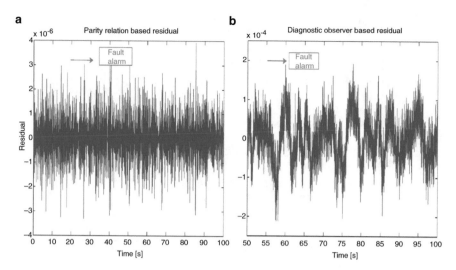

Fig. 4.14 Residuals obtained with (**a**) Parity Space and (**b**) Diagnostic Observers; under the conditions of no faults affecting the system ($f_u = 0$), and up to 8% noise level in the output

In Fig. 4.14 are shown the residuals for the situation when the system is not affected by faults, but the output is corrupted with up to 8% noise level. In Fig. 4.14b a zoom on the residual obtained by the Diagnostic Observer is shown. Both approaches, Parity Space and Diagnostic Observers, generate residuals different from zero due to the noise affecting the output of the system. As a consequence, both approaches provide false alarms. This fact is related to the lack of robustness in the implementation of both FDI model based methods applied. In order to increase robustness, some thresholds could be established.

In Fig. 4.15 it is shown the result of the estimation of f_u with FD-IPM when the output is also corrupted with up to 8% noise level. Its estimation is close to zero but not equal. In this case, it is also needed some thresholds, which are represented with red lines, and indicates that the estimation values are within the 0.1% of the maximum value allowed to this fault.

The residual obtained with the Parity Space for the situation when the system is affected by the actuator fault $f_u = 0.9$ happening at $t = 50$ s is shown in Fig. 4.16. A zoom on the residual obtained by the application of the Diagnostic Observer is shown in Fig. 4.17. The residuals exceeded the thresholds values at this time, in both approaches. Thus, it indicates that the system is under a fault f_u. Both approaches were able to detect the fault.

On the other hand, with FD-IPM the fault was diagnosed with an estimated value $\bar{f}_u = 0.8940$, see Fig. 4.18.

- Sensitivity analysis:

The residual when the system is affected by an incipient fault, $f_u = 0.08$, starting at $t = 50$ s is shown in Fig. 4.19. In Fig. 4.19a it is also shown that the Parity Space based residual did not exceed the threshold values adopted, which

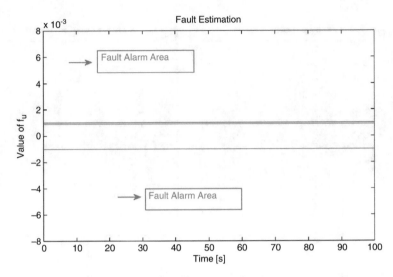

Fig. 4.15 Diagnosis obtained with FD-IPM under the conditions of no faults affecting the system ($f_u = 0$), and up to 8% noise level in the output. The horizontal red lines indicate the thresholds adopted

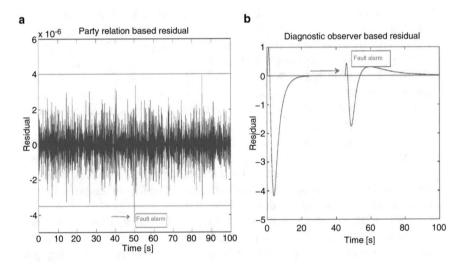

Fig. 4.16 Residuals obtained with (**a**) Parity Space and (**b**) Diagnostic Observers, under the conditions of actuator fault affecting the system ($f_u = 0.9$), and up to 8% noise level in the output. The horizontal red lines indicate the thresholds adopted

are needed for a robust diagnosis. Thus, for Parity Space the thresholds that avoid false alarms do not allow detecting incipient faults. In Fig. 4.19b it is shown that Diagnostic Observers detected the fault. A zoom on the residual obtained by the Diagnostic Observer is shown in Fig. 4.20.

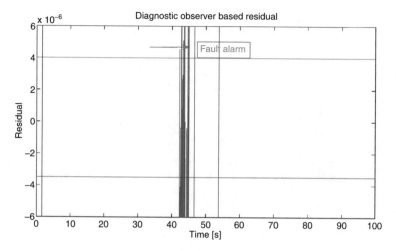

Fig. 4.17 Zoom on the residual presented in Fig. 4.16b, obtained with the application of Diagnostic Observers. The horizontal red lines indicate the thresholds adopted

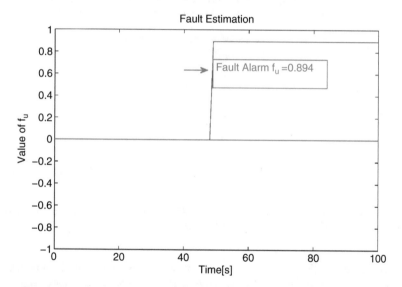

Fig. 4.18 Diagnosis obtained with FD-IPM, under the conditions of actuator fault affecting the system ($f_u = 0.9$), and up to 8% noise level in the output

The fault is also detected with the PSO-M application. Its estimated value is $f_u = 0.08$, see Fig. 4.21. The same thresholds established in order to obtain robustness were used for obtaining sensitivity.

A few remarks regarding the computational effort are necessary. The main disadvantage of FD-IPM, when compared with the Parity Space and Diagnostic Observers approach, is the computational time required. Once the parity vector

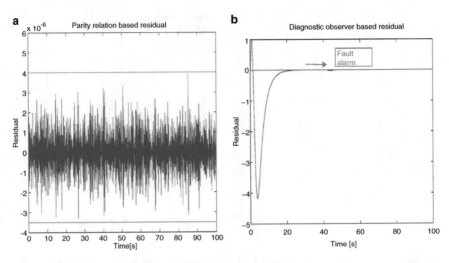

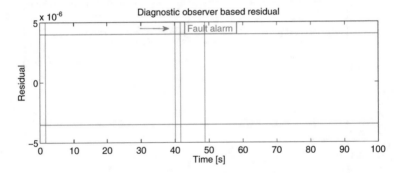

Fig. 4.19 Residuals obtained with (**a**) Parity Space and (**b**) Diagnostic Observers, under the conditions of an actuator incipient fault affecting the system ($f_u = 0.08$) and up to 8% noise level affecting the output. The horizontal red lines indicate the thresholds adopted

Fig. 4.20 Zoom on the residual presented in Fig. 4.19b obtained with the application of Diagnostic Observers. The horizontal red lines indicate the thresholds adopted

is computed, the Parity Space approach only needs matrix multiplications, while the diagnosis based on FD-IPM needs to perform many simulations of the model of the system. For the cases whose results are presented in Figs. 4.16 and 4.19, the diagnosis based on FD-IPM required 906 (174 s) and 1081 (305 s), respectively. On the other hand, Diagnostic Observers took around 16 s for computing the residual. It must be emphasized though that most industrial processes have high time constants, and therefore the processing time required by FD-IPM poses no impairment regarding its application in practical situations.

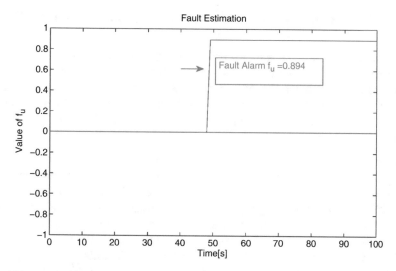

Fig. 4.21 Diagnosis obtained with FD-IPM, under the conditions of actuator incipient fault affecting the system ($f_u = 0.08$), and up to 8% noise level affecting the system. The horizontal red lines indicate the thresholds adopted

4.2.6 Conclusions

The experiments performed indicate that the diagnosis based on FD-IPM is feasible.

The study of the influence of the PSO and ACO parameters permitted the analysis of the influence of diversification and intensification in the quality of the diagnosis. The manipulation of these parameters in order to increase robustness and sensitivity avoided the efforts demanded by the robust residual generation in other model-based FDI methods. The PSO and ACO variants that gave the best results for each algorithm were those that allowed a higher diversification during the search.

The study performed revealed that an adequate balance between diversification and intensification is essential for the development of a robust and sensitive diagnosis based on FD-IPM.

After the comparison of FD-IPM with other two model-based methods, Parity Space and Diagnostic Observers, other advantages of this proposal were identified: it permits to diagnose the system even when the number of faults is higher than the number of sensors; it allowed to obtain robust and sensitive diagnosis in a simple way; it is a general methodology that does not depend on the type of model used to describe the system; no further elements need to be designed or computed, and only the simulations of the model of the system are required.

The application of the Wilcoxon's test indicated that PSO-M over performed PSO_{I1} taking the computational cost as metrics, while maintaining the quality of the estimates obtained.

4.3 Experiments with the Inverted Pendulum System (IPS)

It was shown in Sect. 2.5 that for the IPS, as described in Sect. 2.4, only the detection is possible (Step 2 of FD-IPM). This section shows the application of Step 2 for detecting the faults in the IPS. Moreover, the Step 3 of the FD-IPM was applied, in order to confirm the conclusion after the verification of hypothesis H3 of FD-IPM in Chap. 2: the faults affecting the system can only be detected. The experiments were divided into four parts:

- *Part 1: Analysis of the diversification influence in the diagnosis*
  Designed for analyzing the influence in the diagnosis of the diversification in the ACO and ACO-d search procedures. The variants $ACO_{1.1}$, $ACO_{1.2}$, and $ACO_{1.3}$, which were also applied to the experiments with the DC Motor, see Table 4.4, were applied. The difference between these variants is in the value of the parameter q_0 which influences the trend of the search, because it handles its balance between diversification and intensification. The variants of ACO-d are in correspondence with the variants of ACO, taking $C_{dis} = 0.10$ in all cases. For this part, Cases 1 and 2 from Table 4.12 were considered. These cases represent faulty situations for the simplified IPS, which can be diagnosed based on FD-IPM. The output of the system is corrupted with up to 5% noise level.
- *Part 2: Detecting faults with Fault Diagnosis-Inverse Problem Methodology (FD-IPM)*
  In order to test the results of the Structural Analysis of Fault Diagnosis Inverse Problem for the IPS obtained in Sect. 2.5, which allowed to verify hypothesis H3 of FD-IPM, this part includes experiments with the IPS with no simplifications, i.e. the IPS cannot be diagnosed. As shown in Table 4.12, Case 3 represents a multiple fault situation and Case 4 a situation with no faults. The application of the FI-IPM Step 2 for the detection and the influence of the f_{umbral} in the quality of the result of the detection are presented. For these cases the system output is also corrupted with up to 5% noise level.
- *Part 3: Robustness analysis*

Table 4.12 Faulty situations considered for the experiments with the Inverted Pendulum System (ISP)

Case	f_u	$f_{y(1)}$	$f_{y(2)}$	Noise (%)
1	0.5	-	-	5
2	0.5	-	0.02	5
3	0.5	0.01	0.02	5
4	0	0	0	5
5	0.5	-	-	8
6	0.5	-	0.02	8
7	0.5	0.01	-	8
8	-	0.01	0.02	8
9	0.012	-	-	8
10	0.012	-	0.005	8

The situations under study are Cases 5–8 from Table 4.12, corresponding also to the simplified version of the IPS, which can be diagnosed. The output of the system is corrupted with up to 8% noise level. The variants used for ACO and ACO-d were selected taking into account the results in the previous part of the experiments with IPS.

- *Part 4: Sensitivity analysis*

   The situations under study are Cases 9 and 10 from Table 4.12, corresponding also to the simplified version of the IPS, which can be diagnosed. The output of the system is corrupted with up to 8% noise. The variants of ACO and ACO-d algorithms are taken as in the robustness analysis.

In order to obtain statistically valid conclusions, each algorithm was run 30 times for each faulty situation under study. The notations used in tables and figures coincide with the one introduced in Sect. 4.2: $\bar{f}(\hat{\boldsymbol{\theta}}_f)$ for the mean value of the objective function; $\overline{Eval}$ for the mean of the objective function number of evaluations performed until the minimum value of the objective function is achieved; $\bar{f}_u$, $\bar{f}_{y(1)}$, and $\bar{f}_{y(2)}$ for the mean values of the estimates for the faults f_u, $f_{y(1)}$, and $f_{y(2)}$, respectively. It was also computed the variance of the fault estimated values. This statistics gives a measure of dispersion of the estimates from its mean. For denoting the variance of each fault estimation, the following notation is also introduced: $\sigma^2_{\bar{f}_u}$, $\sigma^2_{\bar{f}_{y(1)}}$, and $\sigma^2_{\bar{f}_{y(2)}}$. The symbols $\overline{Iter}$ and $\bar{t}$ were introduced for denoting mean of the number of iterations and mean of CPU time, respectively, until the stopping criteria are reached. The analysis of the computational cost was based on $\overline{Iter}$, by considering that all the runs were made in a same computer.

In Figs. 4.22 and 4.23 the output of the IPS, in its original version, for two different faulty situations is presented.

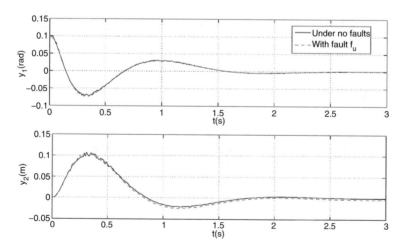

Fig. 4.22 Comparison of the output of the system when the system is not affected by faults, and when the system is affected by an actuator fault f_u, and considered up to 5% noise level

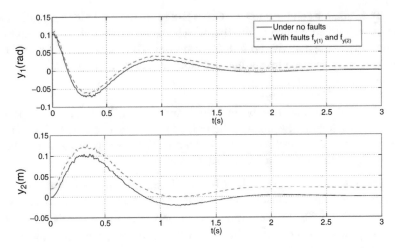

Fig. 4.23 Comparison of the output of the system when the system is not affected by faults, and when the system is affected by two sensor faults $f_{y(1)}$ and $f_{y(2)}$, and considered up to 5% noise level

Table 4.13 Parameter values used for the ACO and ACO-d algorithms in the experiments with the IPS

Notation	K	q_0	Z	C_{inc}	C_{evap}
$ACO_{1.1}$, ACO-d1	63	0.15	30	0.30	0.10
$ACO_{1.2}$, ACO-d2	63	0.55	30	0.30	0.10
$ACO_{1.3}$, ACO-d3	63	0.85	30	0.30	0.10

A performance comparison between ACO and ACO-d is made, using the value of q_0 that was selected in the first part of the experiments with the IPS. The conclusions of the comparison are based on Wilcoxon's test [30]. The criteria for the comparison were: mean value of the objective function $\bar{f}(\hat{\boldsymbol{\theta}}_f)$, and mean number of iterations that were performed until the value of the objective function did not improve, $\overline{Iter}$.

4.3.1   Implementations

- Implementation of ACO and ACO-d

   The implementations were based on the algorithms presented in Algorithm 5. In Table 4.13 are shown the parameter values used for each ACO and ACO-d variant. For ACO-d it was also considered $C_{dis} = 0.10$ for all cases.
- Stopping Criteria

The stopping criteria used for the ACO and ACO-d algorithms were:

- Criterion 1: Maximum number of iterations: $MaxIter = 100$.
- Criterion 2: Maximum number of iterations for which the best value of the objective function remains constant: $Itr_{cte} = 10$
- Criterion 3: Value of the objective function $f(\hat{\boldsymbol{\theta}}_f) \leq 0.01 y_{max}^2$, being y_{max} the maximum value of all the measured output of the system during the considered sampling time.

4.3.2 Results of the Diagnosis with Ant Colony Optimization and Ant Colony Optimization with Dispersion

4.3.2.1 Part 1: Analysis of the Diversification Influence in the Diagnosis

As shown in Sect.2.5, in the structural analysis for the IPS benchmark problem, see Chap. 2, the three faults can only be detected (FD-IPM Steps 1 and 2) because there is no separability among them with the two sensors considered (the system does not satisfy hypothesis H3 due to the non-separability of the faults, but the detectability condition is met).

It was also concluded in Chap. 2 that when considering in the IPS model that the system cannot be affected by more than two faults (a simplification of the described IPS benchmark problem), then the system can be diagnosed with the FD-IPM. Therefore, it has only sense the analysis of the performance of metaheuristics for this simplification of the IPS. The numerical results presented in Table 4.14 are related to the diagnosis of Cases 1 and 2 listed in Table 4.12. These two cases consider that the IPS cannot be affected by more than two faults (the system satisfies H3). In Case 1 there is only one fault, in the actuator, while in Case 2 two faults are considered, one in the actuator and the other in the sensor that measures the position of the car. The bold values are used to emphasize the best numerical value obtained at each Case 1 and 2, respectively. The results show that the diagnosis of the IPS simplified problem based on FD-IPM is possible for these two cases. That corroborates the conclusion achieved in the IPS structural analysis performed in Sect.2.5. The best estimated values for the faults, with the smallest variance, the lowest number of iterations, and the lowest computational time, were obtained in both cases, 1 and 2, with ACO-d1. The variant $ACO_{1.1}$ gave the second better results for both cases.

In Fig. 4.24 are graphically shown the mean values and standard deviations for the fault estimates in Cases 1 and 2 from Table 4.12. The standard deviation is one descriptive statistics for the estimates dispersion. It is computed as $\sigma = \sqrt{\sigma^2}$, and it has the same dimension as the mean value. Therefore it can be used for the estimate dispersion in the same graphic with the mean value of the fault estimates.

Table 4.14 Results for the fault diagnosis faults in Cases 1 and 2 in Table 4.12 with ACO and ACO-d variants listed in Table 4.13, using data up to 5% noise level

Case	Variant	$\bar{f}_u$	$\bar{f}_{y(1)}$	$\bar{f}_{y(2)}$	$\sigma^2_{\bar{f}_u}$	$\sigma^2_{\bar{f}_{y(1)}}$	$\sigma^2_{\bar{f}_{y(2)}}$	$\bar{t}(sec)$	$\overline{Iter}$
$(f_u = 0.5)$ (–) (–)									
1	$ACO_{1.1}$	0.4900	–	–	$1.2 \cdot 10^{-6}$	–	–	35.007	32
	$ACO_{1.2}$	0.4755	–	–	$2.1 \cdot 10^{-5}$	–	–	50.620	48
	$ACO_{1.3}$	0.4606	–	–	$1.5 \cdot 10^{-5}$	–	–	45.421	41
	ACO-d1	**0.4980**	–	–	$\mathbf{1.0 \cdot 10^{-6}}$	–	–	**34.655**	**30**
	ACO-d2	0.4795	–	–	$4.2 \cdot 10^{-5}$	–	–	39.032	38
	ACO-d3	0.4723	–	–	$3.1 \cdot 10^{-5}$	–	–	48.530	47
$(f_u = 0.5)$ (–) $(f_{y(2)} = 0.02)$									
2	$ACO_{1.1}$	0.4891	–	0.0190	$9.3 \cdot 10^{-6}$	–	$1.4 \cdot 10^{-8}$	46.154	43
	$ACO_{1.2}$	0.4701	–	0.0188	$7.4 \cdot 10^{-5}$	–	$1.0 \cdot 10^{-7}$	59.962	57
	$ACO_{1.3}$	0.4499	–	0.0185	$7.6 \cdot 10^{-5}$	–	$3.1 \cdot 10^{-7}$	60.005	58
	ACO-d1	**0.4958**	–	**0.0198**	$8.1 \cdot 10^{-6}$	–	$\mathbf{1.2 \cdot 10^{-8}}$	**45.094**	**42**
	ACO-d2	0.4797	–	0.0182	$7.4 \cdot 10^{-5}$	–	$2.8 \cdot 10^{-7}$	57.082	55
	ACO-d3	0.4555	–	0.0170	$8.0 \cdot 10^{-5}$	–	$9.0 \cdot 10^{-8}$	61.992	60

In Fig. 4.24 are then represented the mean values, and the range with ne standard deviation above and below, i.e. $\bar{f}_u \pm \sigma^2_{\bar{f}_u}$, $\bar{f}_{y(1)} \pm \sigma^2_{\bar{f}_{y(1)}}$, and $\bar{f}_{y(2)} \pm \sigma^2_{\bar{f}_{y(2)}}$. The best fault estimates value, with the smaller standard deviation, are obtained in both Cases 1 and 2, with ACO-d1.

In Fig. 4.25 are represented the accumulative relative percentage errors obtained with the variants of ACO and ACO-d, when estimating faults in Cases 1 and 2 in Table 4.12. It can be observed that ACO provided a higher accumulative error for Case 1. For Case 2, ACO shows a higher accumulative error for f_u, and ACO-d for $f_{y(2)}$. It can also be observed that for ACO and ACO-d, the smaller relative error is always obtained with the variant that considers $q_0 = 0.15$ ($ACO_{1.1}$ and ACO-d1 in Table 4.13 both indicated in blue in Fig. 4.25). Therefore, it can be concluded that the best value for q_0 is 0.15 which corresponds, as showed in Sect. 4.2, to a higher diversification of the search.

Figure 4.26 is similar to Fig. 4.25, but it shows the results from Table 4.14 regarding the computational effort, represented by the average number of iterations required, $\overline{Iter}$. The best results are also obtained with the variants $ACO_{1.1}$ and ACO-d1. Furthermore, the three variants of ACO-d achieved a lower sum of the mean value of iterations than the three variants of ACO. It can be concluded, as in the DC Motor benchmark problem, that a higher diversification improves the diagnosis results. For the experiments that will be presented next, it is only considered the variants $ACO_{1.1}$ and ACO-d1.

Fig. 4.24 Mean value and standard deviation representation for the fault estimates in Cases 1 and 2 from Table 4.12, using ACO and ACO-d variants, and with up to 5% noise level. (**a**) Case 1, $f_u = 0.5$. (**b**) Case 2, $f_u = 0.5$. (**c**) Case 2, $f_{y(2)} = 0.02$

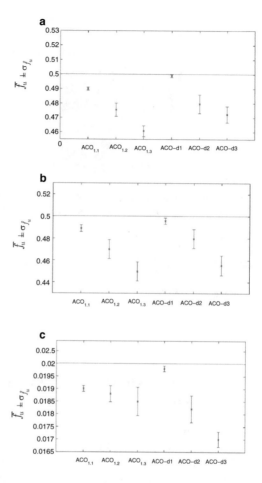

Fig. 4.25 Accumulative relative percentage errors for the fault estimates in Cases 1 and 2 from Table 4.12, using ACO and ACO-d variants and data with up to 5% noise level

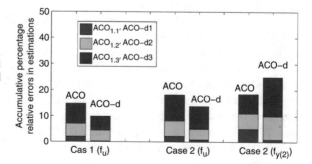

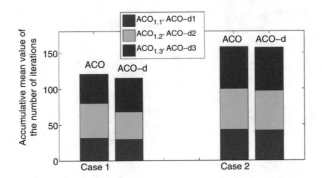

Fig. 4.26 Accumulative mean value of iterations required by ACO and ACO-d for the fault diagnosis in Cases 1 and 2 from Table 4.12, using data with up to 5% noise level

Table 4.15 Results for the fault diagnosis with $ACO_{1.1}$ and ACO-d1 variants, for the Cases 3 and 4 in Table 4.12, and using data with up to 5% noise level

Case	Variant	$\bar{f}_u$	$\bar{f}_{y(1)}$	$\bar{f}_{y(2)}$
($f_u = 0.5$) ($f_{y(1)} = 0.01$) ($f_{y(2)} = 0.02$)				
3	$ACO_{1.1}$	0.3892	0.007	0.017
	ACO-d1	0.0013	0.0001	0.0173
($f_u = 0$) ($f_{y(1)} = 0$) ($f_{y(2)} = 0$)				
4	$ACO_{1.1}$	0.2197	0.01	0.016
	ACO-d1	0.3540	0.0093	0.02

4.3.2.2   Part 2: Detecting Faults with Fault Diagnosis: Inverse Problem Methodology

The objective here is to show the importance of the verification of hypothesis H3, in order to know the limitations of the diagnosis based on FD-IPM, as well as the application of Step 2 of FD-IPM for detecting faults. For that reason, Cases 3 and 4 from Table 4.12 are considered. For these two cases, the original IPS, and not its simplified version, is considered, i.e. the three faults f_u, $f_{y)1)}$, and $f_{y)2)}$ are represented in the model that describes the system. The results of the structural analysis for the IPS indicate, see Sect. 2.5, that in these cases the faults affecting the system can only be detected. It means that the Step 2 of FD-IPM can be applied, but there is no sense in applying Steps 3 and 4 to these cases. However, Step 3 was applied in order to corroborate this conclusion. It is reasonable to expect that the fault estimates obtained for these cases will not be good. Table 4.15 shows the results of the fault estimations for Cases 3 and 4 with the application of the $ACO_{1.1}$ and ACO-d1 variants. As expected from the results of the structural analysis of the IPS, the mean value of the fault estimates in these cases are very different from their real values:

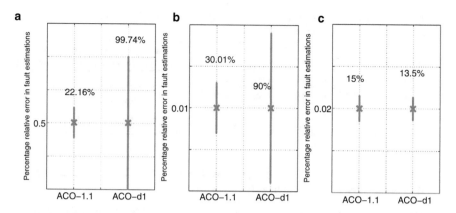

Fig. 4.27 Percentage relative error for the estimates obtained with $ACO_{1.1}$ and ACO-d1 when diagnosing faults in Case 3 from Table 4.12, and using data with up to 5%. (**a**) $f_u = 0.5$. (**b**) $f_{y(1)} = 0.01$. (**c**) $f_{y(2)} = 0.02$

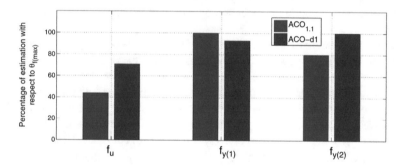

Fig. 4.28 Percentage that represents the fraction of each estimated value obtained with $ACO_{1.1}$ and ACO-d1 when diagnosing Case 4 from Table 4.12, with respect to its maximum permissible value $\theta_{f(\max)} = [0.5\ \ 0.01\ \ 0.02]^T$, and using data with up to 5% noise level

- Case 3: If the diagnosis is based on the fault estimated values for Case 3, then a wrong diagnosis is made. In Fig. 4.27 it is represented the relative percentage error for each estimation, i.e. $\left|\frac{f_u - \hat{f}_u}{f_u}\right| 100$, $\left|\frac{f_{y(1)} - \hat{f}_{y(1)}}{f_{y(1)}}\right| 100$, and $\left|\frac{f_{y(2)} - \hat{f}_{y(2)}}{f_{y(2)}}\right| 100$, in Case 3. The percentage relative errors obtained with ACO-d1 are higher than 90% for two out of the three faults.
- Case 4: In Fig. 4.28 it is shown which percentage represents the value of each fault estimate for Case 4, with respect to its maximum permissible value $\theta_{f(\max)} = [0.5\ \ 0.01\ \ 0.02]^T$ (see restrictions in the optimization problem for the IPS described in Eq. (2.37)). The smaller percentage relative error was observed for the estimation of f_u with the $ACO_{1.1}$ variant: 43.94%. In two cases the fault estimations coincide with the maximum permissible value for the fault. Considering that the system was not under the action of faults, the diagnosis

Table 4.16 Results for the application of the FD-IPM Step 2, in Cases 3 and 4 from Table 4.12, using data with up to 5% noise level

Case	f_{umbral}	$f(0)$	Result	Maximum relative error between measured and computed output (%)
3	0.0003	1.7	The system is affected by faults	1
4	0.00025	0.008	The system is affected by faults	1
3	0.0026	1.7	The system is affected by faults	8
4	0.0020	0.008	The system is not affected by faults	8

based on these estimations will provoke false alarms. As a consequence, the system diagnosis is not satisfactory.

The results of the experiments for Cases 3 and 4, described in Table 4.12, confirm the importance of performing the structural analysis for verifying hypothesis H3 of FD-IPM, in order to know the application limits of the diagnosis based on this methodology. In these cases the system does not satisfy hypothesis H3, for this reason, it is only possible the detection by means of the application of the FD-IPM Step 2.

For the application of Step 2, a value for the f_{umbral} is needed. This value can be determined following the ideas for computing f_{stop} in the stopping criterion number 3 of the metaheuristics, as shown in Sect. 4.1, that will be described next. Let's consider, as in the computation of f_{stop}, that the relative error between the measured and computed output is less than 1% and the total number of sampling time $I = 100$. In that case, $f_{umbral} = 0.01 y_{max}^2$. For Cases 3 and 4 it is obtained that $y_{max} = 0.1788$ and $y_{max} = 0.15876$, respectively. In the fourth column of Table 4.16, $f(0)$ is the value of the objective function, $f(\hat{\theta}_f)$, considering that there are no faults, i.e. $\hat{\theta}_f = 0$. Therefore, when $f(0) < f_{umbral}$, it is considered that system is not affected by faults. The results of the application of Step 2 for $f_{umbral} = 0.01 y_{max}^2$ are showed in the first two lines of Table 4.16 is the value The results of the application of Step 2 with the considered f_{umbral} indicate that in both cases the system is affected by faults. This is a good result for Case 3, but it is false for Case 4. The explanation to the wrong result for Case 4 can be found in the value of f_{umbral}, which was computed considering that the relative error between the measured and computed output of the system is less than 1%. But in both cases the noise affecting the system is up to 5%. Therefore, the relative error between the measured and computed output should be greater than 1%. Considering a f_{umbral} based on a small relative error between the measured and computed output does not give a wrong result of the detection if the system is affected by faults (this is the situation in Case 3), but it is problematic if the system is not affected by faults (this is the situation in Case 4). The last two lines of Table 4.16 also show the results of the application of Step 2 to Cases 3 and 4, but considering a new f_{umbral}. This time the value for f_{umbral} was determined considering that the relative error between the measured and computed output of the system is less than 8%, i.e $f_{umbral} = 0.08 y_{max}^2$. As the third and fourth lines of Table 4.16 show, the results are in both cases right. This example shows how a previous knowledge about the level noise affecting the system can be introduced in f_{umbral} in order to improve the reliability of the diagnosis.

4.3.2.3 Part 3: Robustness Analysis

The robustness analysis is based on experiments performed with Cases 5–8 from Table 4.12, where the high noise level (up to 8%) has the objective of simulating in the output the effect of disturbances affecting the system. In all these cases, the system satisfies hypothesis H3 because it has been considered the simplified version of the IPS benchmark, i.e. no more than two faults can affect it and only two faults are represented in the model of the IPS.

Table 4.17 shows that ACO-d1 over performed $ACO_{1.1}$ in three out of the four cases with respect to the objective function value. The bold values are used to emphasize the best numerical value obtained at each situation represented in this table. Therefore, the fault estimates obtained with ACO-d1 were better in three out of the four cases. Considering the value for $\overline{Iter}$, ACO-d1 is also better in three out of the four cases. In two of the three cases in which ACO-d1 over performed $ACO_{1.1}$ with respect to the objective function, it also occurred with respect to $\overline{Iter}$.

In Fig. 4.29 are shown the percentage relative errors of the fault estimates obtained with $ACO_{1.1}$ and ACO-d1 when diagnosing faults in Cases 5–8 from Table 4.12, using data with up to 8% noise level. For Cases 5, 6, and 8, the percentage relative errors obtained with ACO-d1 are smaller for all the estimations performed. For Case 7, the relative errors obtained with $ACO_{1.1}$ for both faults are smaller, but the maximum relative error obtained with ACO-d1 is not so high as the one obtained for $f_{y(2)}$ in Case 8 with $ACO_{1.1}$. ACO-d1 shows a more uniform behavior when diagnosing different fault situations than $ACO_{1.1}$. The

Table 4.17 Results for the fault diagnosis with $ACO_{1.1}$ and ACO-d1 variants, for the Cases 5–8 from Table 4.12, and using data with up to 8% noise level

Case	Variant	$\bar{f}_u$	$\bar{f}_{y(1)}$	$\bar{f}_{y(2)}$	$\overline{Iter}$	$\bar{f}(\hat{\boldsymbol{\theta}}_f)$
$(f_u = 0.5)\,(-)\,(-)$						
5	$ACO_{1.1}$	0.4863	–	–	43	4.2330
	ACO-d1	**0.4946**	–	–	**41**	**2.7280**
$(f_u = 0.5)\,(-)\,(f_{y(2)} = 0.02)$						
6	$ACO_{1.1}$	0.4788	–	0.0181	**59**	4.4304
	ACO-d1	**0.4936**	–	**0.0192**	62	**4.4119**
$(f_u = 0.5)\,(f_{y(1)} = 0.01)\,(-)$						
7	$ACO_{1.1}$	**0.4932**	**0.0091**	–	73	**2.9860**
	ACO-d1	0.4901	0.0083	–	**60**	3.9746
$(-)\,(f_{y(1)} = 0.01)\,(f_{y(2)} = 0.02)$						
8	$ACO_{1.1}$	–	0.0070	0.00183	78	5.8285
	ACO-d1	–	**0.0083**	**0.0190**	**64**	**4.0975**

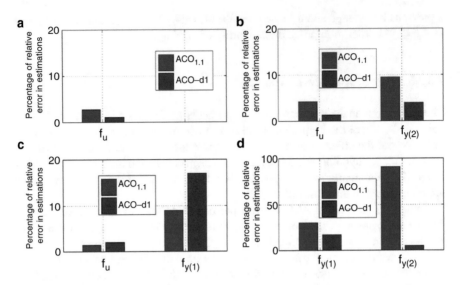

Fig. 4.29 Percentage of relative error for the estimates obtained with the variants $ACO_{1.1}$ and ACO-d1 when diagnosing faults in Cases 5–8 from Table 4.12, using data with up to 8% noise level. (**a**) Cases 5. (**b**) Cases 6. (**c**) Cases 7. (**d**) Cases 8

smaller percentage relative errors obtained with ACO-d1 could be explained, as observed in the results of the experiments with the DC Motor, by the fact that ACO-d1 performs a higher diversification of the search than $ACO_{1.1}$.

4.3.2.4   Part 4: Sensitivity Analysis

The sensitivity analysis that will be presented next for the IPS is based on experiments performed with Cases 9 and 10 in Table 4.12, which consider incipient faults (faults of small magnitude), and a highly noisy environment (up to 8% noise level). In all these cases, the system satisfies hypothesis H3 because it has been considered the simplified version of the IPS benchmark, i.e. the model of the IPS considers that only two faults can affect the system.

Table 4.18 shows that ACO-d1 over performed $ACO_{1.1}$ in both Cases 9 and 10. As occurred with the results presented in the previous section, related to Part 3, it indicates that a higher diversification of the search is better for a more sensitive diagnosis. In this table the bold values are also used to emphasize the best numerical value obtained at each situation.

Table 4.18 Results for the fault diagnosis with $ACO_{1.1}$ and ACO-d1 variants, for Cases 9 and 10 from Table 4.12, and using data with up to 8%

Case	Variant	$\bar{f}_u$	$\bar{f}_{y(1)}$	$\bar{f}_{y(2)}$	$\overline{Iter}$	$\bar{f}(\hat{\theta}_f)$
$(f_u = 0.012)$ (–) (–)						
9	$ACO_{1.1}$	0.010	–	–	100	6.1390
	ACO-d1	**0.011**	-	-	**95**	**4.1216**
$(f_u = 0.012)$ (–) $(f_{y(2)} = 0.005)$						
10	$ACO_{1.1}$	0.0098	–	0.0040	100	6.066
	ACO-d1	**0.014**	–	**0.0046**	**95**	**4.8123**

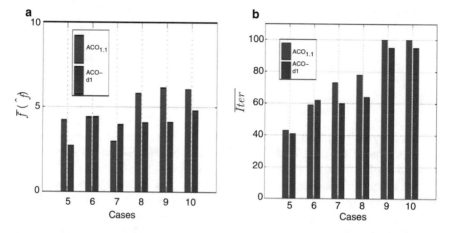

Fig. 4.30 Performance comparison for $ACO_{1.1}$ and ACO-d1 when diagnosing faults in Cases 5–10 from Table 4.12, using data with up to 8% noise level. (**a**) Mean value of the objective function. (**b**) Mean value of the number of iterations

4.3.3   Comparison Between Ant Colony Optimization and Ant Colony Optimization with Dispersion

In the first part of the experiments, $ACO_{1.1}$ and ACO-d1 were selected as the best variants of ACO and ACO-d. Their performances were compared when diagnosing faults in the cases that considered the simplified version of the ISP, i.e. Cases 1 and 2 and Cases 5–10 from Table 4.12.

In Fig. 4.30 it is shown a graphical comparison for the results of the Fault Diagnosis in Cases 5–10 with the variants $ACO_{1.1}$ and ACO-d1.

The results obtained with the application of Wilcoxon's test for the variants $ACO_{1.1}$ and ACO-d1 performance comparison are shown in Table 4.19. The results indicate that ACO-d1 shows a significant improvement over the estimations obtained by $ACO_{1.1}$ with a lower number of iterations, with a significance level $\alpha = 0.1$. In other words, ACO-d1 permits to obtain more accurate estimates, yielding a more reliable diagnosis, and at a lower computational effort.

Table 4.19 Wilcoxon's test results: ACO-d1 versus ACO$_{1.1}$

Comparison	Criterion	R^+	R^-	$T = \min\{R^+, R^-\}$	Critical value of W	α
ACO-d1 vs ACO$_{1.1}$	$\bar{f}(\hat{\theta}_f)$	31	5	5	6	0.1
ACO-d1 vs ACO$_{1.1}$	$\overline{Iter}$	32	4	4	6	0.1

Table 4.20 Faulty situations considered for the first part of experiments (Part 1—Diagnosis with ACO and DE) with the Two Tanks System

Case	$f_{p(1)}$	$f_{p(2)}$	Noise (%)
1	0.2	0.2	5
2	0.05	0	5
3	0.6	0	8
4	0.6	0.6	8
5	0	0	8
6	0.05	0.05	8

4.3.4   Conclusions

The study performed with the ISP benchmark problem also indicates that the fault diagnosis based on the FD-IPM is feasible. FD-IPM also permitted the direct estimation of faults in the simplified version of IPS.

The study of the influence of the search performed by the algorithms ACO and ACO-d in the desired characteristics of fault diagnosis indicated that a greater diversification of the search is important for a sensitive and robust diagnosis. Based on the results obtained for the experiments considered, it is evident that a better diagnosis was performed with ACO-d1. This is due to the fact that the ACO-d conception is based on a pheromone dispersion that helps to enhance the diversification of the search. This is confirmed with statistical tests.

The results based on structural analysis for verifying hypothesis H3 provide prior information for determining the limitations of the diagnosis based on FD-IPM. That was confirmed for the cases of study. The application of Step 2 allows to detect if the IPS, in its general (original) version is under the effect of faults.

4.4   Experiments with the Two Tanks System

As described in Sect. 2.5.3, the Two Tanks System satisfies hypothesis H1-H3 of the FD-IPM. The algorithms ACO, DE, PCA, and DEwPC were applied in the FD-IPM Step 3, i.e. they were applied in order to solve the optimization problem, see Eq. (2.66), that results after the application of the first step of the methodology FD-IPM to the Two Tanks System, which was described in Sect. 2.4.3. The experiments for this benchmark were divided into two parts:

- *Part 1: Diagnosis with ACO and DE*

  Designed for the analysis of the diagnosis based on ACO and DE, in order to obtain a hybrid strategy designed to improve their performances. The situations under study are presented in Table 4.20. This Part 1 is also divided into two parts:

Table 4.21 Faulty situations considered for the second part of experiments (Part 2—Diagnosis with PCA and DE) with the Two Tanks System

Case	$f_{p(1)}$	$f_{p(2)}$	Noise (%)
7	0.25	0.25	2
8	0.25	0.25	5
9	0.08	0	8
10	0	0	8
11	0.02	0.02	8

- General analysis

    It is designed for the general analysis of the diagnosis based on ACO and DE. The situations under study are Cases 1 and 2 from Table 4.20 using data with up to 5% noise level.
- Performance comparison for ACO, DE, and ACO-DE

    It is designed for comparing the diagnosis based on ACO, DE and the hybrid algorithm ACO-DE (described in Sect. 3.4.3). The situations under study are Cases 3–6 from Table 4.20, which consider a noise level in the output up to 8%.

- *Part 2: Diagnosis with PCA and DE*

    It is designed for the analysis of the diagnosis based on PCA and DE, with the aim of obtaining a new algorithm which improves their performances (see algorithm DEwPC in Sect. 3.6). The situations to be studied are presented in Table 4.21.

    The experiments are divided into two parts:

- General and robustness analysis

    It is designed for the analysis of the diagnosis based on PCA and DE. The situations under study are Cases 7 and 8 from Table 4.21.
- Sensitivity analysis and comparison with DEwPC

    It is designed for the sensitivity analysis of the diagnosis with DE, PCA, and DEwPC. The situations under study are Cases 9–11 from Table 4.21. These cases represent situations of incipient faults (Cases 9 and 11), and situations free of faults (Cases 10). The noise level in the output in Cases 9–11 is up to 8 %.

In order to obtain statistically valid conclusions, each algorithm was run 30 times for each faulty situation under study. The notation used in the following tables and figures used to present the diagnosis results are compatible with the one introduced in Sects. 4.2 and 4.3: $\bar{f}_{p(i)}$ and $\sigma^2_{f_{p(i)}}$ correspond to the mean value and variance, respectively, of the estimates for fault $f_{p(i)}$. The computational time was computed during the experiments in order to evaluate the feasibility of the application of the algorithms in real time diagnosis.

Just to give an example of the time response for the Two Tanks System, in Fig. 4.31 are shown the input (inflow to the tanks) and the output (level of the tanks) for one faulty situation considered in the simulations.

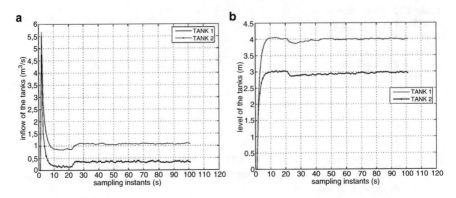

Fig. 4.31 Closed loop behavior of the Two Tanks System when affected by both process faults, $f_{p(1)} = f_{p(2)} = 0.2$, and affected by up to 2–5% noise level. (**a**) Input. (**b**) Output

4.4.1 Implementations

- Implementation of ACO

  In this case the influence of ACO's parameters in the diagnosis was not analyzed. The parameters were set as: $Z = 20$ (ten times the number of faults to be diagnosed); $q_0 = 0.85$; $K = 127$; $C_{inc} = 0.30$ and $C_{evap} = 0.10$. These values correspond to the variant $ACO_{2.3}$, see Table 4.4, which was also applied to the experiments with the DC Motor benchmark problem. The implementation was based on the algorithm presented in Algorithm 5.

- Implementation of DE

  The DE implementation was based on the algorithm shown in Algorithm 1. The parameters were set as $Z = 20$; $C_{cross} = 0.9$ and $C_{scal} = 0.5$.

- Implementation of PCA

  The PCA implementation was based on the algorithm shown in Algorithm 4. The parameter was set as: $MaxIter_c = 10$.

- Implementation of ACO-DE

  The implementation was based on the description of the hybrid strategy made in Sect. 3.4.3. The parameters were set as described next: (1) for ACO, $Z = 20$, $q_0 = 0.55$, $K = 63$, $C_{inc} = 0.30$ and $C_{evap} = 0.10$; (2) for DE, $Z = 10$, $C_{cross} = 0.7$ and $C_{scal} = 0.6$.

- Implementation of DEwPC

  The implementation was based on the algorithm shown in Algorithm 7. Following the recommended parameter values for DEwPC (see Table 3.2 in Sect. 3.6), in the experiments performed its parameters were set as: $Z = 10$; $C_{cross} = 0.9$; $C_{scal} = 0.5$ and $MaxIter_c = 2$.

- Stopping Criteria

  The stopping criteria for ACO, DE, PCA, and DEwPC were:

  – Criterion 1: Maximum number of iterations: $MaxIter = 100$.

- Criterion 2: Maximum number of iterations for which the best value of the objective function remains constant (does not improve): $Itr_{cte} = 10$.
- Criterion 3: Value of the objective function $f(\hat{\boldsymbol{\theta}}_f) \leq 0.01 y_{max}^2$, being y_{max} the maximum value of all the measured output of the Two Tanks System during the considered sampling time.

The stopping criteria for ACO-DE were:

- Criterion 1: Maximum number of iterations for ACO and DE: $MaxIter_{ACO} = 30$ and $MaxIter_{DE} = 70$.
- Criterion 2: Maximum number of iterations for which the best value of the objective function remains constant for DE (does not improve): $Itr_{cte} = 10$.
- Criterion 3: Value of the objective function for DE $f(\hat{\boldsymbol{\theta}}_f) \leq 0.01 y_{max}^2$, being y_{max} the maximum value of all the measured output of the Two Tanks System during the considered sampling time.

4.4.2 Results of Part 1: Diagnosis with Ant Colony Optimization and Differential Evolution

4.4.2.1 General Analysis

Table 4.22 shows the results of the diagnosis with ACO and DE for Cases 1 and 2 from Table 4.20. In this table the bold values are also used to emphasize the best numerical value obtained between ACO and DE at each case. Case 1 represents a situation where both faults are affecting the system, and have the same value. In Case 2, only one fault is affecting the system, $f_{p(1)}$. Both cases consider a noise level up to 5%. The fault estimates provided by DE are better in both cases considering the mean value of the objective function. The number of the objective function evaluations is higher for DE in both cases, but its computational time is smaller than the one needed for ACO, even for the latter performing a smaller number of function evaluations.

In Fig. 4.32 it is shown the sum of the absolute error for both fault estimates obtained with DE and ACO, i.e. $\Delta \bar{f}_T = \Delta \bar{f}_{p(1)} + \Delta \bar{f}_{p(2)}$, when diagnosing Cases 1 and 2 from Table 4.20, being $\Delta \bar{f}_{p(1)} = |\bar{f}_{p(1)} - f_{p(1)}|$ and $\Delta \bar{f}_{p(2)} =$

Table 4.22 Results for the Fault Diagnosis with the algorithms ACO and DE for the Cases 1 and 2 from Table 4.20, using data with up to 5% level noise

Alg	Case	$\bar{f}_{p(1)}$	$\bar{f}_{p(2)}$	$\overline{Eval}$	$\bar{t}(s)$
Case	Alg	($f_{p(1)} = 0.2$)	($f_{p(2)} = 0.2$)		
1	DE	**0.1983**	**0.2001**	1342	**68.774**
	ACO	0.1801	0.2114	**1118**	102.0047
($f_{p(1)} = 0.05$)	($f_{p(2)} = 0$)				
2	DE	**0.05082**	**0.00016**	1168	**64.858**
	ACO	0.06676	0.00032	**664**	78.1

$|\bar{f}_{p(2)} - f_{p(2)}|$. It was not considered the relative error of the estimates for the comparison purpose because the value of both faults to be diagnosed is the same for Case 1, which means that the absolute error can be used as the metrics for comparing the quality of the estimations. For Case 2, one of the faults to be estimated is zero ($f_{p(2)} = 0$), which does not make feasible the use of the relative error, and the other which is different from zero ($f_{p(1)} = 0.05$) gave absolute errors in the order of the absolute error of $f_{p(2)}$). Therefore, the absolute error can be used for the quality of the estimations comparison. In Fig. 4.32 it is shown that for both Cases 1 and 2, DE provides the smallest sum for the absolute error. Moreover, in each case the sum of the absolute error for both fault estimates obtained with DE is smaller than the absolute error obtained with ACO for the $f_{p(1)}$ fault estimate.

In Fig. 4.33 it is represented the mean value of the objective function at each iteration, $\bar{f}(\hat{\boldsymbol{\theta}}_f)$, in order to show the evolution of both algorithms, when diagnosing faults in Case 1 from Table 4.20. It is observed that ACO does not provide a

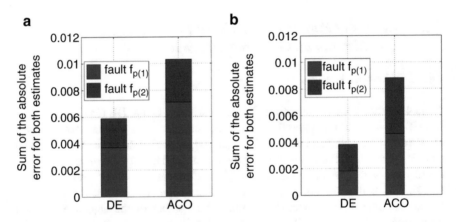

Fig. 4.32 Sum of the absolute error for both fault estimates obtained with DE and ACO, for Cases 1 and 2 from Table 4.20, and using data with up to 5% noise level. (**a**) Cases 1. (**b**) Cases 2

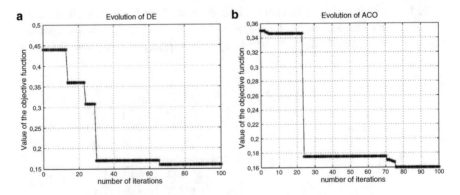

Fig. 4.33 (**a**) Evolution of DE and (**b**) ACO, when diagnosing Case 1 from Table 4.20

significant improvement of the objective function after iteration $Iter = 25$. This fact is related to the value of the parameter $q_0 = 0.85$, which does not permit a higher diversification during the search. In its turn, DE seems to be very dependent on the initial population solution candidates, i.e. the quality of its result improves with a better initial population. This result suggests a convenient way to hybridize both algorithms, see hybrid strategy ACO-DE in Sect. 3.4.3. In such strategy it is applied ACO with a relative small value for K (it is used $K = 63$), for reducing the possible values of the variables, and therefore, the search space. For ACO it is also considered the value $q_0 = 0.55$, which keeps a balance between the intensification and the diversification of the search. Subsequently, it is used the better ants of the history of ACO (in this case the best ten ants will be used, i.e. the 50% of the population of ants), for the initial population of DE with the aim of performing an intensification around a promising area obtained with ACO. The parameters considered for DE in the hybrid are $C_{\text{cross}} = 0.7$ and $C_{\text{scal}} = 0.6$, which determine a greater intensification of the search.

4.4.2.2 Comparison Among Ant Colony Optimization, Differential Evolution and ACO-DE

Table 4.23 shows the numerical results for the Fault Diagnosis in Cases 3 and 4 from Table 4.20, using DE, ACO, and ACO-DE. In this table the bold values are also used to emphasize the best numerical value obtained between DE, ACO and ACO-DE at each case. Case 3 represents a situation in which only one fault, $f_{p(1)}$, affects the system, while in Case 4 multiple faults are present (both faults $f_{p(1)}$ and $f_{p(2)}$ are affecting the system). Both cases consider a noise level up to 8% in order to analyze the robustness. The best results concerning quality of the estimations and number of evaluations of the objective function are obtained with the hybrid strategy ACO-DE.

In Table 4.24 are shown the best and the worst results obtained with each algorithm, when diagnosing faults in Case 5 from Table 4.20, taking as the comparison metrics the sum of the absolute error for both fault estimates. In this table the bold values are also used to emphasize the best numerical value obtained at each situation. Case 5 represents a situation in which the system is free of faults, but the noise level is up to 8%. It can be observed that ACO-DE reduces the number of

Table 4.23 Performance comparison of the algorithms DE, ACO, and ACO-DE when diagnosing faults in Cases 3 and 4 from Table 4.20, using data with up to 8% noise level

Alg	Case	$\bar{f}_{p(1)}$	$\bar{f}_{p(2)}$	$Eval$
\multicolumn{5}{l}{$(f_{p(1)} = 0.6)\ (f_{p(2)} = 0)$}				
3	DE	0.5459	0.0000	1532
	ACO	0.5038	0.0009	1020
	ACO-DE	**0.6081**	**0.0000**	**530**
\multicolumn{5}{l}{$(f_{p(1)} = 0.6)\ (f_{p(2)} = 0.6)$}				
4	DE	0.6068	0.6109	1018
	ACO	0.5901	0.4683	1020
	ACO-DE	**0.6001**	**0.6013**	**440**

Table 4.24 Best and worst results for the Fault Diagnosis with the algorithms DE, ACO, and ACO-DE for the Case 5 from Table 4.20, using data with up to 8% noise level

Alg	Type of Result	$\hat{f}_{p(1)}$	$\hat{f}_{p(2)}$	$Eval$	$t(s)$
$(f_{p(1)} = 0)\,(f_{p(2)} = 0)$					
DE	Best	**0**	**0**	1240	32.684
ACO	Best	0.0600	0.0900	500	29.865
ACO-DE	Best	**0**	**0**	**140**	**13.929**
DE	Worst	0.1191	0	2000	**135.13**
ACO	Worst	0.1900	0.0300	2000	222.06
ACO-DE	Worst	**0**	**0**	**710**	142.04

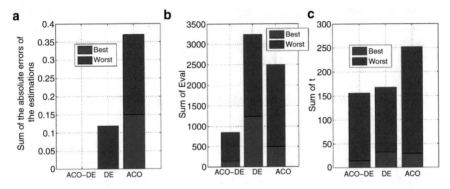

Fig. 4.34 Performance comparison for the algorithms DE, ACO, and ACO-DE when diagnosing faults in Case 5 from Table 4.20, using data with up to 8% noise level. (**a**) Absolute error. (**b**) Eval. (**c**) $t(s)$

objective function evaluations: its worst result is better than the best results obtained with DE and ACO.

In Fig. 4.34 are graphically shown the results from Table 4.24. In Fig. 4.34a it is shown the sum of the absolute errors of the estimates for the best and the worst performance of each algorithm. The estimations performed with ACO-DE in both cases were successful: the sum of the absolute errors of the estimates provided by ACO-DE in both cases (best and worst) is equal to zero. In Fig. 4.34b it is shown the sum of the number of objective function evaluations for both best and worst performances of each algorithm. The sum of the number of objective function evaluations required by the hybrid strategy ACO-DE, in its best and worst cases, is smaller than the one required in the best case of DE, and it is less than half of the sum of the worst and best cases of ACO. In Fig. 4.34c it is shown the sum of the computational time for the best and the worst performances of each algorithm. The sum of the computational time required by ACO-DE in both cases, i.e. best and worst, is less than the sum for the other two algorithms DE and ACO.

Table 4.25 Best results for the Fault Diagnosis with the algorithms DE, ACO, and ACO-DE for the Case 6 from Table 4.20, using data with up to 8% noise level

Alg	$\hat{f}_{p(1)}$	$\hat{f}_{p(2)}$	Eval
	$(f_{p(1)} = 0.05)$	$(f_{p(2)} = 0.05)$	
ACO-DE	**0.0646**	**0.051**	220
DE	0.0690	0.0681	1360
ACO	0.07900	0.0900	740

Table 4.26 Nature of iterations in the hybrid strategy ACO-DE: mean value of the number of iterations required by each method ($\overline{Iter}_{ACO}$ and $\overline{Iter}_{DE}$) when diagnosing faults in Cases 3 and 4 from Table 4.20

Case	$\overline{Iter}_{ACO}$	$\overline{Iter}_{DE}$
3	16	37
4	20	22

Table 4.27 Nature of iterations in the hybrid strategy ACO-DE: number of iterations required by each method ($Iter_{ACO}$ and $Iter_{DE}$) for the best and the worst performances when diagnosing faults in Cases 5 and 6 from Table 4.20

Case	Type of results	$Iter_{ACO}$	$Iter_{DE}$
5	Best	10	4
	Worst	30	41
6	Best	8	14

Table 4.25 presents the best results for the Fault Diagnosis in Case 6 from Table 4.20, taking as the comparison metrics the sum of the absolute error of both fault estimates. In this table the bold values are also used to emphasize the best numerical value obtained for each fault. Case 6 represents a situation of multiple incipient faults, with a higher noise level (up to 8%). This situation is for analyzing the sensitivity of the diagnosis. The results show that ACO-DE yields again the best results.

Table 4.26 is intended to show the nature of iterations of ACO-DE. Cases 3 and 4 from Table 4.20 were chosen for such purpose. Each row in Table 4.26 indicates the mean value of the number of iterations executed by ACO and DE, in the hybrid ACO-DE. Table 4.27 has the same purpose of Table 4.26, but it shows the number of iterations for the best and the worst performances of ACO-DE when diagnosing faults in Cases 5 and 6 from Table 4.20.

In Fig. 4.35 are graphically shown the results from Tables 4.26 and 4.27. In Fig. 4.35a it is shown the mean value of the number of iterations required by ACO and DE in the hybrid ACO-DE, for Cases 3 and 4: both algorithms do not reach their maximum prescribed number of iterations in the hybrid strategy ($MaxIter_{ACO} = 30$ and $MaxIter_{DE} = 70$). From Fig. 4.35b it can be observed that ACO-DE only reached the maximum number of iterations for ACO ($MaxIter_{ACO} = 30$) in the worst run for Case 5.

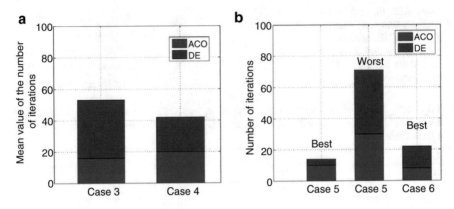

Fig. 4.35 Nature of iterations in the hybrid strategy ACO-DE when diagnosing faults in Cases 3–6 from Table 4.20, using data with up to 8% noise level. (**a**) Mean value of the number of iterations. (**b**) Number of iterations

Table 4.28 Results for the Fault Diagnosis with algorithms PCA and DE for the Cases 7 and 8 from Table 4.21, using data with up to 2% and 5% noise level, respectively

Case	Alg	$\bar{f}_{p(1)}$	$\bar{f}_{p(2)}$	$\sigma^2_{\hat{f}_{p(1)}}$	$\sigma^2_{\hat{f}_{p(2)}}$	$\overline{Eval}$
$(f_{p(1)} = 0.25)$ $(f_{p(2)} = 0.25)$						
7	DE	**0.2463**	**0.2478**	**2.092**·10^{-4}	5.05·10^{-5}	1260
	PCA	0.2429	0.2532	2.473·10^{-4}	**4.513**·10^{-4}	**250**
$(f_{p(1)} = 0.25)$ $(f_{p(2)} = 0.25)$						
8	DE	**0.2482**	**0.2480**	**2.468**·10^{-4}	**3.096**·10^{-4}	1440
	PCA	0.2454	0.2458	3.764·10^{-4}	4.359·10^{-4}	**309**

4.4.3   Results of the Part 2: Diagnosis with Particle Collision Algorithm and Differential Evolution

4.4.3.1   General and Robustness Analysis

Table 4.28 shows the numerical results for the experiments of the second part related to the Two Tanks System. The cases under study are Cases 7 and 8 from Table 4.21, which consider the same magnitude for both faults $f_{(p(1))}$ and $f_{(p(2))}$, but different levels of noise (up to 2% and 5%, respectively). In this table the bold values are also used to emphasize the best numerical value obtained for each fault at each Case 7 and 8, respectively. In Figs. 4.36 and 4.37, the results presented in Table 4.28 are graphically shown. From Fig. 4.36 it can be observed that DE provides better fault estimates with smaller values for the standard deviation ($\sigma = \sqrt{\sigma^2}$) than those obtained with PCA. From Fig. 4.37 it can be observed that the mean value for the objective function evaluations performed by DE is almost five times higher than the number of evaluations performed by PCA. Based on results, it is logic to think

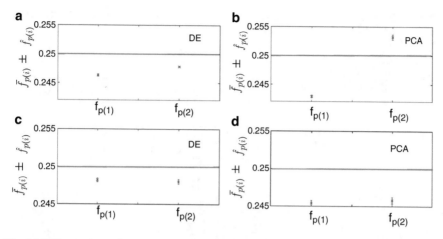

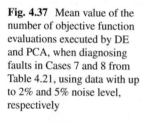

Fig. 4.36 Mean value and standard deviation representation for the fault estimates obtained with DE and PCA in Cases 7 and 8 from Table 4.21, using data with up to 2% and 5% noise level, respectively. The red lines indicate the exact fault values. (**a**) Cases 7. (**b**) Cases 7. (**c**) Cases 8. (**d**) Cases 8

Fig. 4.37 Mean value of the number of objective function evaluations executed by DE and PCA, when diagnosing faults in Cases 7 and 8 from Table 4.21, using data with up to 2% and 5% noise level, respectively

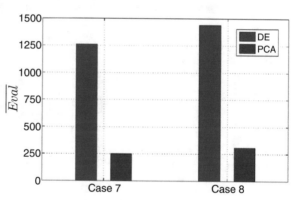

about redesigning DE in order to keep the quality of its estimations, while reducing its computational cost. This was the inspiration for the algorithm DEwPC described in Sect. 3.6.

4.4.3.2 Sensitivity Analysis, Comparison Among Differential Evolution, Particle Collision Algorithm and Differential Evolution with Particle Collision

Table 4.29 shows the numerical results for the Fault Diagnosis in Cases 9 and 10 from Table 4.21, using DE and PCA. Case 9 represents a situation in which one fault is incipient, $f_{p(1)}$, and the other is not affecting the system, $f_{p(2)} = 0$, while Case 10 represents a situation free of faults, i.e. $f_{p(1)} = f_{p(2)} = 0$, but for both

Table 4.29 Results for the Fault Diagnosis with the algorithms DE and PCA for the Cases 9 and 10 from Table 4.21, using data with up to 8% noise level

Case	Alg	$\bar{f}_{p(1)}$	$\bar{f}_{p(2)}$	$\sigma^2_{\hat{f}_{p1}}$	$\sigma^2_{\hat{f}_{p1}}$	$\overline{Eval}$
$(f_{p(1)} = 0.08)\,(f_{p(2)} = 0)$						
9	DE0.0760	**0.0010**	**4.847·10⁻⁵**	**4.2·10⁻⁷**	1880	
	PCA	0.0688	0.0022	1.417·10⁻⁴	1.9·10⁻⁶	**911**
$(f_{p(1)} = 0)\,(f_{p(2)} = 0)$						
10	DE	**0.00060**	**0.00050**	1.818·10⁻⁷	**1.489·10⁻⁷**	1780
	PCA	0.00072	0.00059	**1.689·10⁻⁷**	1.966·10⁻⁷	**659**

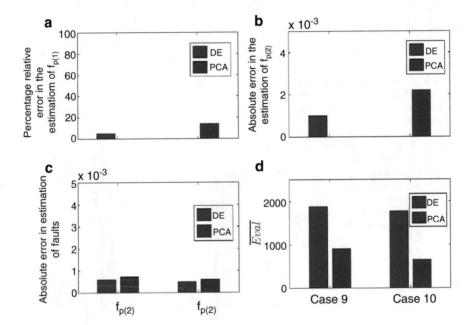

Fig. 4.38 Performance comparison for the algorithms DE and PCA when diagnosing faults in Cases 9 and 10 from Table 4.21, using data with up to 8% noise level. (**a**) Cases 9. (**b**) Cases 9. (**c**) Cases 10. (**d**) $\overline{Eval}$

cases high noise levels are considered, up to 8%, in order to analyze the sensitivity. In this table the bold values are also used to emphasize the best numerical value obtained for each fault at each Case 9 and 10, respectively. In Fig. 4.38 the results are graphically shown. In Figs. 4.38a and b, the results related to the error in the fault estimates for Case 9 are shown. Figure 4.38a shows the percentage relative error obtained in the estimation of $f_{p(1)}$, and Fig. 4.38b shows the absolute error in the estimation of $f_{p(2)}$. For $f_{p(2)}$ it is presented the absolute error because the real value is $f_{p(2)} = 0$. For both faults, the errors obtained with PCA are more than twice the errors provided due to the application of the algorithm DE. Figure 4.38c shows the absolute error for the estimates in Case 10. In this case both real fault

Table 4.30 Best and worst results for the Fault Diagnosis with the algorithms DE, PCA, and DEwPC for the Cases 9 and 11 from Table 4.21, using data with up to 8% noise level

Case	Alg	Type of result	$\hat{f}_{p(1)}$	$\hat{f}_{p(2)}$	$Eval$
$(f_{p(1)} = 0)\ (f_{p(2)} = 0)$					
9	DE	Best	0.0003	0.0080	1240
	PCA	Best	0.0009	0.0085	770
	DEwPC	Best	**0.0000**	**0.0000**	400
	DE	Worst	0.1191	**0.0050**	2000
	PCA	Worst	0.0183	0.0486	1100
	DEwPC	Worst	**0.0150**	0.0196	600
$(f_{p(1)} = 0.02)\ (f_{p(2)} = 0.02)$					
10	DE	Best	0.0221	0.0250	5114
	PCA	Best	0.0253	0.0285	880
	DEwPC	Best	**0.0203**	**0.0201**	455
	DE	Worst	0.0517	0.0421	2000
	PCA	Worst	0.0626	0.0455	1100
	DEwPC	Worst	**0.0222**	**0.0268**	900

values to be estimated are $f_{p(1)} = f_{p(2)} = 0$, and for that reason it is considered the absolute error in the comparison. From Fig. 4.38c it can be observed that DE gives better estimates, but for both algorithms the absolute error values for each fault are below 10_{-5}. Figure 4.38d shows that DE executes in both Cases 9 and 10 more than twice the number of function evaluations than PCA executes. As in the robustness analysis presented in Sect. 4.4.3.1, it is concluded that a redesign of DE based on PCA, seeking to decrease its computational cost, could be interesting.

The new algorithm, based on DE and PCA, is denominated Differential Evolution with Particle Collision (DEwPC), and has been described in Sect. 3.6. In Table 4.30 are shown the numerical results of the comparison among the algorithms DE, PCA, and DEwPC when diagnosing faults in Cases 10 and 11 from Table 4.21. Case 11 represents a situation of incipient faults with a high noise level. In this table the bold values are also used to emphasize the best numerical value obtained for each fault at each Case 10 and 11, respectively. In the worst results, DE and PCA are able to detect the faults, but they overestimate their values, which can lead to wrong decisions. The estimations with DEwPC are more accurate.

Figure 4.39 graphically represents the best and the worst results obtained when diagnosing faults in Case 10 with DE, PCA, and DEwPC. In Fig. 4.39a it is shown the sum of the absolute error of the fault estimates for the best and worst results. It can be observed that the errors for the worst result of the algorithm DEwPC are smaller than the sum of the best and worst errors obtained with DE and PCA. Concerning the computational cost, Fig. 4.39b shows that the sum of the number of objective function evaluations of the best and worst cases for DEwPC is smaller than the number of evaluations that executed DE at its best result, and almost half of the fault estimates absolute error sum for the worst and best results obtained with PCA.

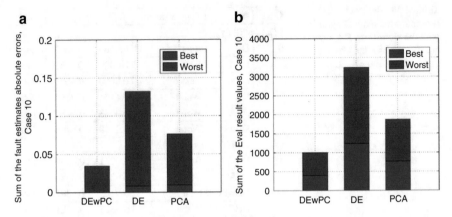

Fig. 4.39 Performance comparison for the algorithms DE, PCA, and DEwPC when diagnosing faults in Case 10 from Table 4.21, using data with up to 8% noise level. (**a**) Absolute errors. (**b**) Eval

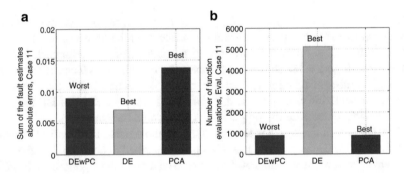

Fig. 4.40 Comparison of the worst performance of DEwPC with the best performances of DE and PCA, when diagnosing faults in Case 11 from Table 4.21, using data with up to 8% noise level. (**a**) Absolute errors. (**b**) Eval

Figure 4.40 graphically represents a comparison between the worst performance of the algorithm DEwPC when diagnosing faults in Case 11 from Table 4.21, and the best performances of DE and PCA for the same case. In Fig. 4.40a it is shown the sum of the absolute error for both fault estimates for the worst performance of DEwPC, and the corresponding best results for DE and PCA. It can be observed that the fault estimates absolute error in the worst result of DEwPC is smaller than the error for the best result for PCA, and only approximately 30% higher than the best performance of DE. Concerning the computational cost, Fig. 4.40b shows that the number of objective function evaluations required by DEwPC at its worst performance is smaller than the corresponding number required by DE, and almost the same with PCA, at their best performances.

4.4.4 Conclusions

The experiments performed and the analysis of their results indicate that the Fault Diagnosis based on FD-IPM is feasible. The analysis of the results also showed how to analyze results related to the performance of metaheuristics, in order to design hybrid strategies or new algorithms with certain goals.

The analysis of the results of the first part of the experiments with the Two Tanks System allowed to propose a hybrid strategy with ACO and DE. The main idea of the proposed hybridization consists of using ACO for obtaining a good initial population for DE. In general the hybrid strategy ACO-DE leads to faster, more robust, and more sensitive (to faults) diagnosis than pure DE or pure ACO.

The general and robustness analysis of the second part of the experiments with the Two Tanks System allowed to think about redesigning DE in order to keep the quality of its estimations, while reducing its computational cost. As a consequence, it was proposed a new algorithm DEwPC which was described and tested in Chap. 3. In the case of the Two Tanks System, the experiments showed that the new algorithm DEwPC provides a better diagnosis than DE and PCA.

4.5 Remarks

This chapter presented the experiments and results related with the application of the third step of the methodology FD-IPM to the three benchmark problems considered, i.e. DC-Motor, Inverted Pendulum System, and Two Tanks System. The third step of FD-IPM consists of the solution of an optimization problem with the use of metaheuristics. The solution of the optimization problem provides estimates for the faults affecting the system. Therefore, the quality of the diagnosis, which is made in step number four of the methodology, is based on the reliability of the estimations obtained in the FD-IPM Step 3.

Experiments were performed for each benchmark problem, and different metaheuristics were applied for the solution of the optimization problems.

The numerical results for the experiments showed the quality of the fault estimates obtained with the use of metaheuristics. Therefore, the results confirmed the feasibility of the diagnosis based on FD-IPM.

The experiments with the DC Motor and the IPS showed how a study of the influence of the metaheuristics parameters permitted the analysis of the influence of diversification and intensification in the quality of the diagnosis. These parameters can be manipulated in order to increase robustness and sensitivity.

The experiments performed with the DC Motor also showed how the analysis of the results of the metaheuristic application can lead to the design of new metaheuristics. In this case the new metaheuristic is denominated PSO-M and was formalized in Chap. 3. In the experiments with the DC Motor, it was also shown a comparison with other well-established model based methods for Fault Diagnosis.

The experiments with the IPS also confirmed that the Structural Analysis, presented in Chap. 2, permits to obtain prior information about the limitations of the diagnosis based on FD-IPM. They also confirmed the importance of verifying hypothesis H3 of FD-IPM.

The experiments with the Two Tanks System allowed to introduce the hybrid strategy ACO-DE, and also the new algorithm denominated DEwPC, which were formalized in Chap. 3.

The results in general showed that a greater diversification in the search is related to higher robustness and sensitivity in the diagnosis. On the other hand, this increment in robustness and sensitivity can provoke also an increase in the computational cost. The experiments showed that the two metaheuristics proposed in this book, DEwPC and PSO-M, and the hybrid strategy ACO-DE, provide an adequate balance between these characteristics.

Chapter 5
Final Remarks

In this book, it is presented the *Fault Diagnosis—Inverse Problem Methodology* (FD-IPM), see Sect. 2.1, which is based on the formulation of FDI as a parameter estimation inverse problem. The FD-IPM consists of three hypotheses and four steps which are extensively described in this book.

The proposed methodology provides a Fault Diagnosis of the system under study, based on numerical estimations of the faults affecting the system. These estimations are the solution of the optimization problem which corresponds to the Fault Diagnosis Inverse Problem, and they are obtained with the use of metaheuristics. Therefore, the quality of the diagnosis is related to the quality of the estimations obtained with the metaheuristics. As a consequence, the selection of a proper metaheuristic is an important task. With the aim of helping to understand the nature or historical roots of the metaheuristics, which can be useful for choosing a metaheuristic, the most common classification features are summarized in a diagram in Fig. 3.1

However, the results of the experiments in Chap. 4, with the application of the methodology to three benchmark problems, show that there is not a general absolute answer to which is the best metaheuristic to be applied. Instead, it is shown that a greater diversification in the search executed by the metaheuristics is related to higher robustness and sensitivity in the diagnosis. An increase in the diversification of the search can be obtained by calibrating some parameters of the metaheuristics. On the other hand, an increase in the diversification can be related to an increase in the computational cost which is related to a higher diagnosing time, then, it is necessary to achieve an adequate compromise between the robustness and sensitivity, and the computational cost. In order to face this problem, it is recommended the combination or hybridization of metaheuristics. In this book are proposed two new metaheuristics, DEwPC and PSO-M in Sects. 3.6 and 3.7, respectively, which are based on a proper combination of metaheuristics, as well as a hybrid strategy ACO-DE in Sect. 3.4.3. The experiments in Chap. 4 have also shown that the two metaheuristics proposed in this book, DEwPC and PSO-M, and

© Springer International Publishing AG, part of Springer Nature 2019
L. Camps Echevarría et al., *Fault Diagnosis Inverse Problems: Solution with Metaheuristics*, Studies in Computational Intelligence 763,
https://doi.org/10.1007/978-3-319-89978-7_5

the hybrid strategy ACO-DE, provide an adequate balance between the desirable characteristics in FDI: robustness, sensitivity, and computational cost.

Summarizing, we would like to list two final important remarks concerning FD-IPM:

- The verification of hypothesis H3, see Sect. 2.1, is very important in order to obtain prior information about the limitations of the diagnosis based on FD-IPM.
- For solving the optimization problem, which is formulated in Step 1 of FD-IPM, see Sect. 2.1, it is recommended the use of combinations or hybridizations of metaheuristics.

5.1   Future Trends

As already mentioned, there is not the single best metaheuristic for solving the optimization problem formulated in Step 1 of FD-IPM. However, for real applications, it could be very useful to provide those responsible for executing the FDI based on FD-IPM with some recommended metaheuristics. For that reason, we consider that future works with the aim of improving the practical applicability of the FD-IPM should intend to develop proposals for collecting information regarding the quality of the diagnosis obtained with certain metaheuristics. This information could be used for building decision tables which can be presented to the experts. Based on these tables, the experts can choose a metaheuristic from the ones recommended, which better satisfies their priorities concerning the diagnosing of the system.

On the other hand, it is interesting to extend the FD-IPM to time variable faults, which can eventually increase the spectrum of applications of this methodology.

Appendix A
Implementation in Matlab® of Differential Evolution with Particle Collision (DEwPC)

A.1 Main Function: DEwPC

```
function [Errorfinal, Eval]=DEwPC(n,z,Cscal,Ccross,
          a,b,... MaxIter,MaxIterc)
% Main function: Algorithm DEwPC
% Author: L\'{\i}dice Camps Echevarr\'{\i}a
% May-2017

% % Input / Entrada:
% n: dimension of the search space (number of
     variables)
% z: number of solutions in the population
% Cscal: Scaling Factor
% Ccross: Crossover Factor
% a,b: minimal and maximum allowed values for the
     variables
%      (It is consideredthe same cotes for each
    variable)
% MaxIter: maximum number of iterations of the main
%          loop of the algorithm
% MaxIterc: maximum number of iterations  to be
%           performed for the operator Local

% % Output /Salida
% ErrorFinal: final error achieved by the algorithm
% Eval: number of final function evaluations performed
%       by the algorithm
```

© Springer International Publishing AG, part of Springer Nature 2019
L. Camps Echevarría et al., *Fault Diagnosis Inverse Problems: Solution with Metaheuristics*, Studies in Computational Intelligence 763,
https://doi.org/10.1007/978-3-319-89978-7

```
% Choosing he function to minimize
%%%%%%%%%%%%%%%%%%%%%%%%%%%%%%%%%%%%%%%

    strfitnessfct ='f2';
    funcMin=1;

%%%%%%%%%%%%%%%%%%
% Stop Criteria
%%%%%%%%%%%%%%%%%%


  ErrorTarget=0.00000001;

 EvalMax=10000*n;

%%% DEwPC parameters and inicialization %%%
    %%%%%%%%%%%%%%%%%%%%%%%%%%%%%%%%%%%%%%%%%%%%%%%%%
    % Initial Population matrix (X0) (1xk+1) (ACO)
    %%%%%%%%%%%%%%%%%%%%%%%%%%%%%%%%%%%%%%%%%%%%%%%%%

    X0= a + (b-a).*rand(n,z);


    %%%%%%%%%%%%%%%%%%%%%%%%%%%%%%%%%%%%%%%%%%%%%%%%%%%
    %  Computing the remaining initial conditions for
        PSO
    %%%%%%%%%%%%%%%%%%%%%%%%%%%%%%%%%%%%%%%%%%%%%%%%%%%

    [Xbest, Errorfinal]=Initio(X0, z, funcMin,
      strfitnessfct);

    Xactual=X0;

    XbestTotal=zeros(n,MaxIter);

    ErrorfinalTotal=zeros(1,MaxIter);

    Eval=z;

    %%%%%%%%%%%%%%%%%%%%%%%%%
    % Main Loop of DEwPC
    %%%%%%%%%%%%%%%%%%%%%%%%%

for i=1:MaxIter

        % Applying Operator Mutation
```

```
    Xtemp=Mutation(Xactual, z, Cscal, Xbest);

% Applying Operator Crossover
    Xtemp=Crossover(Xactual, Xtemp, n, z, Ccross);

    for j=1:z
       FunObj(j)=feval(strfitnessfct, Xactual(:,j));
    end

    % Applying Operator Selection of DEwPC
    for j=1:z

       if j<=sqrt(z)   % Apply Selection from DE

          [Xtemp(:,j), Eval, FunObj(j)]=
                      Selection(Xtemp(:,j),...
                      Xactual(:,j),strfitnessfct,...
                      NumEval, FunObj(j), j);

        else    % Apply    Absorption-Scattering
            from PCA


       [Xtemp(:,j), Eval, FunObj(j)]=
           AbsScatt(Xtemp(:,j),...
           Xactual(:,j), Fbest, MaxIterc,a,b,...
           strfitnessfct, n, j, FunObj(j), NumEval);

        end

     end

    Xactual=Xtemp;

    [Xbest, Errorfinal]=Update(Xactual, funcMin,
       FunObj, z);

Eval=NumEval+(i*z);

    if Eval>=NumEvalMax

      break

    end
```

```
    if Errorfinal<= ErrorTarget

        break

    end

  end
```

A.1.1   Function Initio

```
%%Function Initio
%%This function computes the best initial solution and
%%its value
%Function Name: Initio
%This function returns:
%    Xbest: the best solution of the initial population
%    Errorfinal: the error of the function value at the
%               best solution Xbest

function [Xbest, Errorfinal]=Initio(X0, z, funcMin,...
        strfitnessfct);

F=zeros(z,1);

for i=1:z
    F(i)=feval(strfitnessfct, Xactual(:,i));
end

[Fbest, I]=min(F);

Xbest=Xactual(:,I);

Errorfinal=Fbest-funcMin;
```

A.1.2   Function Mutation

```
%%Function Mutation
%%This function applies Mutation operator to each
    solution
```

```
%%
%Function Name: Mutation
%This function returns:
%    Xtemp: the solutions after the Mutation
%

function Xtemp=Mutation(Xactual, z, Cscal, Xbest)


q= floor(1 + (z).*rand(4,z));

Xtemp=Xactual;

for i=1:z

    Xtemp(:,i)=Xbest+Cscal*(Xactual(:,q(1,i))-...
              Xactual(:,q(2,i))+Xactual(:,q(3,i))-...
              Xactual(:,q(4,i)));

end
```


A.1.3 Function Crossover


```
%%Function Crossover
%%This function applies Crossover operator to each
     solution
%%
%Function Name: Crossover
%This function returns:
%    Xtemp: the solutions after the Crossover
%

function Xtemp=Crossover(Xactual, Xtemp, n, z, Ccross)

q= rand(n,z);

for i=1:z

    for j=1:n

        if q(j,i) > Ccross
```

```
            Xtemp(j,i)=Xactual(j,i);

      end

   end

end
```

A.1.4 Function Selection

```
%%Function Selection
%%This function applies Selection operator to the
    choosen
%%solutions
%Function Name: Selection
%This function returns:
%   Xtemp: the solutions after the Selection
%

function [Xgenerada, NumEval, FunObjj]=Selection
        (Xgenerada,...Xanterior,strfitnessfct,
        NumEval, FunObjj, k)

  FunObjGenerada=feval(strfitnessfct, Xgenerada);

  NumEval=NumEval+1;

  if FunObjGenerada>= FunObjj;

    Xgenerada= Xanterior;

  end
```

A.1.5 Function AbsScatt

```
%%Function AbsScatt
%%This function applies Absorption-Scattering
    operator to
%%the choosen solutions
```

```
%Function Name: AbsScatt
%This function returns:
%    Xtemp: the solutions after the Absorption-
     Scattering
%

function [Xfinal, NumEval, FunObjj]=AbsScatt
            (Xgenerada,... Xanterior, Fbest,MaxIterAbs,
             a, b, strfitnessfct,... n,  k, FunObjj,
             NumEval)

 fa=feval(strfitnessfct,Xgenerada);

 NumEval=NumEval+1

 if fa<= FunObjj

     [Xfinal, FunObjj, NumEval]=Local
                        (Xgenerada,... MaxIterAbs,
                         a, b, n,... strfitnessfct,
                         NumEval);


 else

     if rand>1-(Fbest)/(fa)


         [Xfinal, FunObjj, NumEval] =Local
                            (Xgenerada,... MaxIterAbs,
                             a, b, n,...strfitnessfct,
                             NumEval);

     else

         Xfinal=Xanterior;

     end

 end
```

A.1.6 Function Update

```
%%Function Update
%%This function updates the best values for the new
%%population
%%
%Function Name: Update
%This function returns:
%    Xbest: best solution  from the population
%    Errorfinal: error for the  function value at
%                the best solution Xbest

function [Xbest, Errorfinal]=Update(Xactual,
            funcMin,... FunObj,z)


[Fbesttemp, I]=min(FunObj);

Xbest=Xactual(:,I);

Errorfinal=FunObj(I)-funcMin;
```

A.1.7 Function Local

```
%%Function Local
%%This function makes a search arround a solution
%%
%Function Name: Local
%This function returns:
%    Xbest: best solution  from the population
%    Fbest: function value at the best solution Xbest
%    Errorfinal: error for the  function value at the
%                best solution Xbest
function [Xfinal, temp2FO, Eval]=Local(Xgenerada,...
            MaxIterc,a, b, n, strfitnessfct, NumEval)

Xfinal=Xgenerada;

XfinalTemp=Xfinal;

temp2FO=feval(strfitnessfct, Xfinal);
```

```
for i=1:MaxIterc

    for j=1:n

        q2=0.8+ (1-0.8)*rand;

        q3=1.0+ (1.2-1.0)*rand;

        if q2*Xfinal(j)>=a

            Xlower=q2*Xfinal(j);

        else
            Xlower=a;

        end

         if q3*Xfinal(j)<=b

            Xupper=q3*Xfinal(j);

        else
            Xupper=b;

        end

        q1=rand;

        XfinalTemp(j)=Xfinal(j)+(Xupper-Xfinal(j))*
                      q1-...(Xlower-Xfinal(j))*(1-q1);

    end

    temp1FO=feval(strfitnessfct, XfinalTemp);

    NumEval=NumEval+1;

    if temp1FO <=  temp2FO

        Xfinal= XfinalTemp;

        temp2FO=temp1FO;

    end
end
```

Appendix B
Implementation in Matlab® of Particle Swarm Optimization with Memory (PSO-M)

B.1 Main Function: PSOM

```
function [ErrorFinal, Eval]=PSOM(n,z1,z2,C11,C12,...
        C21,C22,a,b,Cevap,Cinc,k,qo,itr1,itr2)
% Main function: Algorithm PSO-M
% Author: L\'{\i}dice Camps Echevarr\'{\i}a
% May-2017


% % Input / Entrada:
% n: dimension of the search space (number of
%     variables)
% z1: number of particles in the first stage of PSO-M
% z2: number of particles in the second stage of PSO-M
% C11,C12: cognitive and social parameter values
%          in the first stage of PSO
% C21,C22: cognitive and social parameter values in
%          the second stage of PSO-M
% a,b: minimal and maximun allowed values for the
%     variables
%     (It is considered the same cotes for each
%     variable)
% Cevap: evaporation factor
% Cinc: incremental factor
% k: number of possible discrete values for each
%     variable
% qo: control parameter in ACO (level of randomness
%        during the ant generation
% itr1: maximum number of iterations in the first
```

© Springer International Publishing AG, part of Springer Nature 2019
L. Camps Echevarría et al., *Fault Diagnosis Inverse Problems: Solution with Metaheuristics*, Studies in Computational Intelligence 763,
https://doi.org/10.1007/978-3-319-89978-7

```
%          stage of PSO-M
% itr2: maximum number of iterations in the second
%          stage of PSO-M


% % Output /Salida
% ErrorFinal: final error achieved by the algorithm
% Eval: number of final function evaluations performed
%          by the algorithm


% Choosing he function to minimize
%%%%%%%%%%%%%%%%%%%%%%%%%%%%%%%%%%%%%%%%%

    strfitnessfct ='f2';
    funcMin=1;


%%%%%%%%%%%%%%%%%
% Stop Criteria
%%%%%%%%%%%%%%%%%


   ErrorTarget=0.00000001;

 EvalMax=10000*n;

%%% ACO parameters and inicialization %%%
    %%%%%%%%%%%%%%%%%%%%%%%%%%%%%%%%%%%%%%%%%%%%
    % Discreet Values (DV) (1xk+1) (ACO)
    %%%%%%%%%%%%%%%%%%%%%%%%%%%%%%%%%%%%%%%%%%%%

    DV = a:((b-a)/k):b;

    %%%%%%%%%%%%%%%%%%%%%%%%%%%%%%%%%%%%
    % Pheromone matrix  (F) (ACO)
    %%%%%%%%%%%%%%%%%%%%%%%%%%%%%%%%%%%%

    F = 0.50*ones(n,k+1);

%%% PSO inicialization  for the first stage %%%
    %%%%%%%%%%%%%%%%%%%%%%%%%%%%%%%%%%%%%%%%%%%%%%%%%%%%%
    % Initial position of particles in PSO  (X) (PSO)
    %%%%%%%%%%%%%%%%%%%%%%%%%%%%%%%%%%%%%%%%%%%%%%%%%%%%

        X = a + (b-a).*rand(n,z1);
```

```
%%%%%%%%%%%%%%%%%%%%%%%%%%%%%%%%%%%%%%%%
% Initial  pi  of particles in PSO
%%%%%%%%%%%%%%%%%%%%%%%%%%%%%%%%%%%%%%%%

   Pi = X;

%%%%%%%%%%%%%%%%%%%%%%%%%%%%%%%%%%%%%%%%%%%%%%%%%
% Initial velocity of particles in PSO (V)
    (nxz)(PSO)
%%%%%%%%%%%%%%%%%%%%%%%%%%%%%%%%%%%%%%%%%%%%%%%%%

   V = 1 + (b-1).*rand(n,z1); % Save velocity value

%%%%%%%%%%%%%%%%%%%%%%%%%%%%%%%%%%%%%%%%%%%%%%%%%%%%%%
%  Computing the remaining initial conditions
      for PSO
%%%%%%%%%%%%%%%%%%%%%%%%%%%%%%%%%%%%%%%%%%%%%%%%%%%%%%

   [Pg, ErrorPg, FunObjPi]=best(X, z1, funcMin,...
                             strfitnessfct);
   ErrorFinal=ErrorPg;


%%%%%%%%%%%%%%%%%%%%%%%%%%%%%%%%%%%%%%
% First Stage of PSO-M (z1,c1,c2)
%%%%%%%%%%%%%%%%%%%%%%%%%%%%%%%%%%%%%%

for itr=1:itr1

   for i=1:z1

      % Updating position and velocity
      [V,X] = UpdateVX(n,C11,C12,X,Pi,Pg,V,itr,i,
         itr1);

   end

   % Updating Pg and Pi

   [Pg,ErrorPg,Pi,FunObjPi]=UpdatePgPi(X,FunObjPi,
                             Pi,... strfitnessfct,z1,
                             funcMin);
   if ErrorPg < ErrorFinal
      ErrorFinal=ErrorPg;
   end
```

```
    if ErrorFinal < ErrorTarget
        break
    end


    %%%%%%%%%%%%%%%%%%%%%%%%%%%%%%%%%
    % Updating pheromone matrix
    %%%%%%%%%%%%%%%%%%%%%%%%%%%%%%%%%

    Ftemp=(1-Cevap)*F;
    Ftemp1=zeros(n,k+1);
    S=zeros(n,1);

    for t=1:n

        r = find( DV >= Pg(t,:));

        tam = size(r);

        if (tam(1,2) == 0)

            S(t,:) = DV(1,k+1);
            Ftemp1(t,k+1) = Cinc*F(t,k+1);

        else

            l = DV(1,r(1,1)) - Pg(t,:);

            m = ((b-a)/k)/2;

            if (l) > (m)

                if r(1,1) == 1

                    S(t,:) = DV(1,r(1,1));
                    Ftemp1(t,r(1,1)) = Cinc*
                        F(t,r(1,1));

                else

                    S(t,:) = DV(1,(r(1,1))-1);
                    Ftemp1(t,(r(1,1)-1)) = Cinc*F(t,
                        (r(1,1)-1));

                end
```

```
                              else

                                  S(t,:) = DV(1,r(1,1));
                                  Ftemp1(t,r(1,1)) = Cinc*F(t,r(1,1));

                              end

                      end

             end

          F=Ftemp+Ftemp1;


end




%%%%%%%%%%%%%%%%%%%%%%%%%%%%%%%%%%%%%%%%%%%%%%%%%%%%%%%
% Initial parameters for the Second Stage of PSO
%%%%%%%%%%%%%%%%%%%%%%%%%%%%%%%%%%%%%%%%%%%%%%%%%%%%%%%


   PC = UpdatePC(F,n,k);

   S = GenerateAnts(n,z2,qo,F,PC,DV);

   X = S;

   V = 1 + (b-1).*rand(n,z2);


   [PgNo, ErrorPgNo, FunObjPi]=best(X, z2, funcMin,...
                               strfitnessfct);



%%%%%%%%%%%%%%%%%%%%%%%%%%%%%%%%%%%%%%%%%%%
% Second Stage of  PSO-M  (z2,c1,c2)
%%%%%%%%%%%%%%%%%%%%%%%%%%%%%%%%%%%%%%%%%%%

for itr=1:itr2
```

```
for i=1:z2

    [V,X] = UpdateVX(n,C21,C22,X,Pi,Pg,V,itr,
             i,itr2);

end
[Pg,ErrorPg,Pi,FunObjPi]=UpdatePgPi(X,FunObjPi,
                          Pi,... strfitnessfct,z2,
                          funcMin);
if ErrorPg < ErrorFinal
    ErrorFinal=ErrorPg;
end

Eval=itr1*z1+z2*itr;

if Eval >= EvalMax
    break
end

if ErrorFinal < ErrorTarget

    break

end

end
```

B.1.1   Function Best

```
%%Function best
%%This function updates the best values for a swarm
%%
%Function Name: best
%This function returns:
%   Pg: best particle  from the swarm
%   ErrorPg: error for the  function value at
%            the best particle Pg
%   FuncObjPi: updated vector with the values of the
%              function at the Pi of each particle
%
```

```
function [Pg, ErrorPg, FuncObjPi]=best(Xactual, z,...
         fmin, strfitnessfct)

FuncObjPi=zeros(z,1);

for i=1:z
    FuncObjPi(i)=feval(strfitnessfct, Xactual(:,i));
end

[Ftemp, I]=min(FuncObjPi);

Pg=Xactual(:,I);

FuncPg= FuncObjPi(I);

ErrorPg=norm(FuncPg-fmin);
```

B.1.2   Function UpdateVX

```
%%Function UpdateVX
%%This function updates the position and velocity
%%of a particle of the swarm
%%
%Function Name: UpdateVX
%This function returns:
%   V: the velocity of a particle
%   X: the position of a particle
%

function [V,X] = UpdateVX(n,C1,C2,X,Pi,Pg,V,itr,...
                         i,iteraciones)

    w=0.9-((0.9-0.4)/iteraciones)*itr;

    V(:,i) = w*V(:,i) + C1*(diag(rand(1,n)))*
            (Pi(:,i)... -X(:,i))+C2*(diag(rand
            (1,n)))*(Pg - X(:,i));

    X(:,i) = X(:,i) + V(:,i);

end
```

B.1.3   Function UpdatePgPi

```
%%
%Function Name: UpdatePgPi
%This function returns:
%   Pg: best particle  from the swarm
%   ErrorPg: error for the  function value at the
%            best particle Pg
%   Pi: position of the ith particle
%   FuncObjPi: updated vector with the values of the
%              function at the Pi of each particle
%

function[Pg,ErrorPg,Pi,FunObjPi]=UpdatePgPi(X,
                             FunObjPi,... Pi,
                             strfitnessfct,z,fmin)


for k=1:z

    temp=feval(strfitnessfct, X(:,k));

    if temp<FunObjPi(k)

        FunObjPi(k)=temp;
        Pi(:,k)=X(:,k);

    end


end

[Fbesttemp, I]=min(FunObjPi);

Pg=X(:,I);

FunObjPg=FunObjPi(I);


ErrorPg=norm(FunObjPg-fmin);
```

B.1.4   Function UpdatePC

```
%%Function UpdatePC
%%This function updates Accumulative Probability
    matrix
%%
%Function Name: UpdatePC
%This function receives three parameters:
%    F: Pheromone Matrix
%    n: Number of discreet variable
%    k: Number of interval
% and returns:
%    PC: Accumulative Probability matrix
%

function [PC] = UpdatePC(F,n,k)

    PC=zeros(n,k+1);

    for i=1:n   % Loop to move through row (Variable)

        for j=1:1:k+1    % Loop to move through col
                         %(Discreet Value)

            PC(i,j) = sum(F(i,1:j))/sum(F(i,:));

        end

    end


end
```

B.1.5   Function GenerateAnts

```
%%Function GenerateAnts
%%This function generates a  population of ants
%%
%Function Name: GenerateAnts
%This function receives seven parameters:
%    n: number of discreet variable
%    k: number of interval
```

```
%    qo: randomness parameter
%    z: number of ants
%    F:pPheromone Matrix
%    DV: discreet Values
%    PC: accumulative Probability matrix
% and returns:
%    S: matrix with a new population of z ants
%

function [S] = GenerateAnts(n,z,qo,F,PC,DV)

    S = ones(n,z);

    for i=1:z % Loop to generate z ants

        for j=1:n % Loop to generate n discreet values
                  % associate with n variables for
                    each ant

            qn = rand;

            if qn < qo

                [C,I] = max(F(j,:)); e
                S(j,i) = DV(1,I);

            else

                c = find(PC(j,:) >= qn);
                S(j,i) = DV(1,c(1,1));

            end
        end
    end

end
```

References

1. Acosta Diaz, C., Camps Echevarria, L., Prieto Moreno, A., Silva Neto, A.J., Llanes-Santiago, O.: A model-based fault diagnosis in a nonlinear bioreactor using an inverse problem approach. Chem. Eng. Res. Des. **114**, 18–29 (2016)
2. Angeli, C., Chatzinikolaou, A.: On-line fault detection techniques for technical systems: a survey. Int. J. Comput. Sci. Appl. **I**(I), 22–30 (2004)
3. Angeline, P.J.: Chapter Evolutionary optimization versus particle swarm optimization: philosophy and performance differences. In: Evolutionary Programming VII: Proceeding of the Seventh Annual Conference on Evolutionary Programming (EP98). Lecture Notes in Computer Science, vol. 1447, pp. 601–611. Springer, New York (1998)
4. Baeck, T.: Evolutionary Algorithms in Theory and Practice: Evolution Strategies, Evolutionary Programming, Genetic Algorithms. Oxford University Press, Oxford (1996)
5. Barbosa Muniz, W., Campos Velho, H.F., Manuel Ramos, F.: A comparison of some inverse methods for estimating the initial condition of the heat equation. J. Comput. Appl. Math. **103**, 145–163 (1999)
6. Bauer, F., Hohage, T., Munk, A.: Iteratively regularized Gauss-Newton method for nonlinear inverse problems with random noise. SIAM J. Numer. Anal. **47**, 1827–1846 (2009)
7. Becceneri, J.C., Zinober, A.: Extraction of energy in a nuclear reactor. In: XXXIII Simposio Brasileiro de Pesquisa Operacional. Campos do Jordão, SP, Brazil (2001)
8. Becceneri, J.C., Sandri, S., Luz, E.F.P.: Using ant colony systems with pheromone dispersion in the traveling salesman problem. In: Proceedings of the 11th International Conference of the Catalan Association for Artificial Intelligence. Sant Martí d'Empúries (2008)
9. Beck, J.V.: Combined parameter and function estimation in heat transfer with application to contact conductance. J. Heat Trans. **110**(4b), 1046–1058 (1998)
10. Beielstein, T., Parsopoulos, K.E., Vrahatis, M.N.: Tuning PSO parameters through sensitivity analysis. Tech. rep., Reihe Computational Intelligence CI 124/02, Collaborative Research Center (SFB 531), Department of Computer Science and University of Dortmund (2002)
11. Beni, G., Wang, J.: Swarm intelligence. In: Seventh Annual Meeting of the Robotics Society of Japan. RSJ Press, Tokyo (1989)
12. Bertsekas, D.P.: Nonlinear Programming. Athena Scientific, Belmont, MA (1999)
13. Blum, C.: Ant colony optimization: introduction and recent trends. Phys. Life Rev. **2**(4), 353–373 (2005)
14. Bonabeau, E., Dorigo, M., Theraulaz, G.: Swarm Intelligence: From Natural to Artificial Systems. Oxford University Press, Oxford (1999)

© Springer International Publishing AG, part of Springer Nature 2019
L. Camps Echevarría et al., *Fault Diagnosis Inverse Problems: Solution with Metaheuristics*, Studies in Computational Intelligence 763,
https://doi.org/10.1007/978-3-319-89978-7

15. Brest, J., Greiner, S., Boscovic, B., Mernik, M., Zumer, V.: Self-adapting control parameters in differential evolution: a comparative study on numerical benchmark problems. IEEE Trans. Evol. Comput. **10**(8), 646–657 (2006)
16. Camacho, O., Padilla, D., Gouveia, J.L.: Fault diagnosis based on multivariate statistical techniques. Rev. Téc. Ing. Univ. Zulia. **30**(3), 253–262 (2007)
17. Campos Velho, H.F.: Inverse Problems in Space Research. SBMAC, Sao Carlos, Brazil (2008)
18. Camps Echevarría, L., Llanes-Santiago, O., Silva Neto, A.J.: Chapter Fault diagnosis in industrial systems using bioinspired cooperative Strategies. In: Nature Inspired Cooperative Strategies for Optimization, NICSO 2010. Studies in Computational Intelligence, vol. 284. Springer, New York (2010)
19. Camps Echevarría, L., Llanes-Santiago, O., Silva Neto, A.J.: A proposal to fault diagnosis in industrial systems using bio-inspired strategies. Ingeniare. Revista chilena de ingeniería **19**(2), 240–252 (2011)
20. Camps Echevarría, L., Silva Neto, A.J., Llanes-Santiago, O., Hernández Fajardo, J.A., Jiménez Sánchez, D.: A variant of the particle swarm optimization for the improvement of fault diagnosis in industrial systems via faults estimation. Eng. Appl. Artif. Intell. **28**, 36–51 (2014)
21. Camps Echevarria, L., Campos Velho, H.F., Becceneri, J.C., Silva Neto, A.J., Llanes-Santiago, O.: The fault diagnosis inverse problem with ant colony optimization and ant colony optimization with dispersion. Appl. Math. Comput. **227**(15), 687–700 (2014)
22. Cao, H., Haiwen, Y., Zhao M. Shi Jian, S., Limei, Z., Xin, G.: The fault diagnosis of aircraft power system based on inverse problem of fuzzy optimization. Part G: J. Aerosp. Eng. **230**(6), 1059–1074 (2016)
23. Carlisle, A., Dozier, G.: An off-the-self PSO. In: Proceedings of the Particle Swarm Optimization Workshop, Indiana, pp. 1–6 (2001)
24. Chen, J., Patton, R.J.: Robust Model-Based Fault Diagnosis for Dynamic Systems. Kluwer Academic Publishers, Dordrecht (1999)
25. Chow, E.Y., Willsky, A.: Analytical redundancy and the design of robust failure detection systems. IEEE Trans. Autom. Control **29**, 603–614 (1984)
26. Clerc, M., Kennedy, J.: The particle swarm-explosion, stability and convergence in a multidimensional complex space. IEEE Trans. Evol. Comput. **6**(2), 58–73 (2002)
27. Das, S., Abraham, A., Uday, K.C., Konar, A.: Differential evolution using a neighborhood-based mutation operator. IEEE Trans. Evol. Comput. **13**(3), 526–553 (2009)
28. Dawkins, R.: The Selfish Gene. Clarendon Press, Oxford (1976)
29. de Miguel, L.J., Blázquez, L.F.: Fuzzy logic-based decision-making for fault diagnosis in a DC motor. Eng. Appl. Artif. Intell. **18**(1), 423–450 (2005)
30. Derrac, J., García, S., Molina, D., Herrera, F.: A practical tutorial on the use of nonparametric statistical test as a methodology for comparing evolutionary and swarm intelligence algorithms. Swarm Evol. Comput. **1**(1), 3–18 (2011)
31. Ding, S.X.: Model-Based Fault Diagnosis Techniques: Design Schemes, Algorithms, and Tools. Springer, Berlin (2008)
32. Dorigo, M.: Ottimizzazione, apprendimento automatico, ed algoritmi basati su metafora naturale. Ph.D. thesis, Politécnico di Milano (1992)
33. Dorigo, M., Blum, C.: Ant colony optimization theory: a survey. Theor. Comput. Sci. **344**(2–3), 243–278 (2005)
34. Dorigo, M., Maniezzo, V., Colorni, A.: The ant system: optimization by a colony of cooperating agents. IEEE Trans. Syst. Man Cybern. Part B **26**(1), 29–41 (1996)
35. Duarte, C., Quiroga, J.: PSO algorithm for parameter identification in a DC motor. Rev. Fac. Ing. Univ. Antioquia **55**, 116–124 (2010). (In Spanish)
36. Dulmage, A., Mendelsohn, N.: Coverings of bipartite graphs. Can. J. Math. **10**(5), 517–534 (1958)
37. Eberhart, R., Shi, Y.: Chapter Comparison between Genetic Algorithms and Particle Swarm Optimization. In: Evolutionary Programming VII: Proceeding of the Seventh Annual Conference on Evolutionary Programming (EP98). Lecture Notes in Computer Science, vol. 1447, pp. 611–619. Springer, Berlin (1998)

38. Eberhart, R.C., Shi, Y.H.: Comparing inertia weights and constriction factors in particle swarm optimization. In: Proceeding of the IEEE Congress on Evolutionary Computation, pp. 84–88 (2001)
39. Engl, H.W., Hanke, M., Neubauer, A.: Regularization of Inverse Problems. Kluwer Academic Publishers, Dordrecht (1996)
40. Fliess, M., Join, C., Mounier, H.: An introduction to nonlinear fault diagnosis with an application to a congested internet router. In: Advances in Communication and Control Networks. Lecture Notes in Control and Information Sciences. Springer, Berlin (2004)
41. Florin Metenidin, M., Witczak, M., Korbicz, J.: A novel genetic programming approach to nonlinear system modelling: application to the DAMADICS benchmark problem. Eng. Appl. Artif. Intell. **17**(4), 363–370 (2004)
42. Frank, P.M.: Fault diagnosis in dynamic systems using analytical and knowledge-based redundancy – a survey and some new results. Automatica **26**(3), 459–474 (1990)
43. Frank, P.M.: Analytical and qualitative model-based fault diagnosis – a survey and some new results. Eur. J. Control **2**(1), 6–28 (1996)
44. Frank, P.M., Koeppen-Selinger, B.: New developments using AI in fault diagnosis. Eng. Appl. Artif. Intell. **10**(1), 3–14 (1997)
45. García, S., Molina, D., Lozano, M., Herrera, F.: A study on the use of non-parametric tests for analyzing the evolutionary algorithms behaviour: a case study on the CEC 2005 Special Session on Real Parameter Optimization. J. Heuristics **15**(6), 617–644 (2009)
46. Glover, F.: Tabu search: a tutorial. Tech. rep., Center for Applied Artificial Intelligence, University of Colorado (1990)
47. Goldberg, D.E.: Genetic Algorithms in Search, Optimization, and Machine Learning. Addison-Wesley, Reading, MA (1989)
48. Gomez, Y.: Two step swarm intelligence to solve the feature selection problem. J. Univ. Comput. Sci. **14**(15), 2582–2596 (2008)
49. Gong, W., Cai, Z., Ling, C.X., Li, H.: A real-coded biogeography-based optimization with mutation. Appl. Math. Comput. **216**(9), 2749–2758 (2010)
50. Gong, W., Cai, Z., Ling, C.X.: DE/BBO a hybrid differential evolution with biogeography-based optimization for global numerical optimization. Soft Comput.: Fusion Found. Methodol. Appl. Arch. **15**(4), 645–665 (2010)
51. Hadamard, J.: Sur les problèmes aux dérivées partielles et leur signification physique, pp. 49–52. Princeton University Bulletin, Princeton, NJ (1902)
52. Hart, W.E., Krasnogor, N., Smith, J.E.: Recent Advances in Memetic Algorithms. Studies in Fuzziness and Soft Computing. Springer (2005). http://books.google.com.br/books?id=LYf7YW4DmkUC
53. Henrique Terra, M., Tinós, T.R.: Fault detection and isolation in robotic manipulators via neural networks: a comparison among three architectures for residual analysis. J. Robot. Syst. **7**(18), 367–374 (2001)
54. Hoefling, T.: Detection of parameter variations by continuous-time parity equations. In: 12th IFACWorld-Congress, pp. 511–516 (1993)
55. Hoefling, T., Isermann, R.: Fault detection based on adaptive parity equations and single-parameter tracking. Control Eng. Pract. **4**(10), 1361–1369 (1996)
56. Hohmann, H.: Automatische ueberwachung und fehlerdiagnose am werkzeugmaschinen. Ph.D. thesis, Technische Hochschule Darmstadt (1987)
57. Isermann, R.: Process fault detection based on modelling and estimation methods – a survey. Automatica **20**(4), 387–404 (1984)
58. Isermann, R.: Fault diagnosis of machines via parameter estimation and knowledge processing. Automatica **29**(4), 815–835 (1993)
59. Isermann, R.: Model based fault detection and diagnosis methods. In: American Control Conference, pp. 1605–1609 (1995)
60. Isermann, R.: Supervision, fault-detection and fault-diagnosis methods- an introduction. Control Eng. Pract. **5**(5), 639–652 (1997)

61. Isermann, R.: Model based fault detection and diagnosis. Status and applications. Annu. Rev. Control **29**(1), 71–85 (2005)
62. Isermann, R.: Fault-Diagnosis Systems: An Introduction from Fault Detection to Fault Tolerance. Springer, Berlin (2006)
63. Isermann, R.: Fault-Diagnosis Applications: Model-Based Condition Monitoring: Actuators, Drives, Machinery, Plants, Sensors, and Fault-tolerant Systems. Springer, Berlin (2011)
64. Isermann, R., Ballé, P.: Trends in the application of model-based fault detection and diagnosis of technical processes. Control Eng. Pract. **5**(5), 709–719 (1997)
65. Kameyama, K.: Particle swarm optimization – a survey. IEICE Trans. Inf. Syst. **E92-D**(7), 1354–1361 (2009)
66. Kanović, Z., Rapaić, M.R., Jeliićc, Z.D.: Generalized particle swarm optimization algorithm – theoretical and empirical analysis with application in fault detection. Appl. Math. Comput. **217**(24), 10175–10186 (2011)
67. Karaboga, D., Akay, B.: A survey: algorithms simulating bee swarm intelligence. Artif. Intell. Rev. **31**(1), 61–85 (2009)
68. Keller, J.B.: Inverse problems. Am. Math. Mon. **83**, 107–118 (1976)
69. Kennedy, J.: The particle swarm: social adaptation of knowledge. In: IEEE International Conference on Evolutionary Computation, pp. 303–308. IEEE, Piscataway, NJ (1997)
70. Kennedy, J.: Chapter The behavior of particles. In: Evolutionary Programming VII: Proceeding of the Seventh Annual Conference on Evolutionary Programming (EP98). Lecture Notes in Computer Science, vol. 1447, pp. 581–590. Springer, New York (1998)
71. Kennedy, J., Eberhart, R.: Particle swarm optimization. In: IEEE International Conference on Neural Networks, vol. 4, pp. 1942–1948. IEEE, Perth (1995)
72. Kiran, M., Ozceylan, E., Gunduz, M., Paksoy, T.: A novel hybrid approach based on particle swarm optimization and ant colony algorithm to forecast energy demand of Turkey. Energy Convers. Manag. **53**(2), 75–83 (2012)
73. Kirkpatrick, S., Gelatt, C.D., Vecchi, M.P.: Optimization by simulated annealing. Science **220**(4598), 671–680 (1983)
74. Knupp, D.C., Silva Neto, A.J., Sacco, W.F.: Estimation of radiative properties with the particle collision algorithm. In: Inverse Problems, Design and Optimization Symposium, Miami, Florida (2007)
75. Krasnogor, N., Smith, J.: A tutorial for competent memetic algorithms: model, taxonomy, and design issues. IEEE Trans. Evol. Comput. **9**(5), 474–488 (2005)
76. Krishnaswami, V., Luh, G.C., Rizzoni, G.: Nonlinear parity equation based residual generation for diagnosis of automotive engine faults. Control Eng. Pract. **3**(10), 1385–1392 (1995)
77. Krysander, M., Aslund, J., Frisk, E.: An efficient algorithm for finding minimal overconstrained subsystems for model-based diagnosis. IEEE Trans. Syst. Man Cybern. Part A: Syst. Hum. **38**(1), 197–206 (2008)
78. Krysander, M., Frisk, E.: Sensor placement for fault diagnosis. IEEE Trans. Syst. Man Cybern. Part A: Syst. Hum. **38**(6), 1398–1410 (2008)
79. Larrañaga, P., Lozano, J.: Estimation of Distribution Algorithms: A New Tool for Evolutionary Computation. Genetic Algorithms and Evolutionary Computation, vol. 2. Kluwer Academic Publishers, Boston (2002)
80. Li, Z., Dahhou, B.: A new fault isolation and identification method for nonlinear dynamic systems: application to a fermentation process. Appl. Math. Model. **32**, 2806–2830 (2008)
81. Liang, J.J., Qin, A.K., Suganthan, P.N., Baskar, S.: Comprehensive learning particle swarm optimizer for global optimization of multimodal functions. IEEE Trans. Evol. Comput. **10**(3), 281–295 (2006)
82. Liu, Q., Wenyuan, L.: The study of fault diagnosis based on particle swarm optimization algorithm. Comput. Informa. Sci. **2**(2), 87–91 (2009)
83. Liu, L., Yang, S., Wang, D.: Particle Swarm Optimization with Composite Particle in Dynamic Environments. IEEE Trans. Syst. Man Cybern. Part B: Cybern. **40**(6), 1634–1638 (2010)

84. Lobato, F.S., Steffen, V., Silva Neto, A.J.: Solution of inverse radiative transfer problems in two-layer participating media with differential evolution. Inverse Prob. Sci. Eng. **18**(2), 183–195 (2009)

85. Lobato, F.S., Steffen, V., Silva Neto, A.J.: Solution of the coupled inverse conduction-radiation problem using multi-objective optimization differential evolution. In: 8th World Congress on Structural and Multidisciplinary Optimization. Lisboa, Portugal (2009)

86. Lunze, J.: Laboratory three tanks system -benchmark for the reconfiguration problem. Tech. rep., Technical University of Hamburg-Harburg, Institute of Control Engineering, Germany (1998)

87. Luz, E.F.P., Becceneri, J.C., De Campos Velho, H.F.F.: A new multiparticle collision algorithm for optimization in a high-performance environment. J. Comput. Interdiscip. Sci. **1**(1), 3–10 (2008)

88. Mauryaa, M.R., Rengaswamy, R., Venkatasubramaniana, V.: Fault diagnosis using dynamic trend analysis: a review and recent developments. Eng. Appl. Artif. Intell. **20**(2), 133–146 (2007)

89. Metenidin, M.F., Witczak, M., Korbicz, J.: A novel genetic programming approach to nonlinear system modelling: application to the DAMADICS benchmark problem. Eng. Appl. Artif. Intell. **24**, 958–967 (2011)

90. Metropolis, N., Rosenbluth, A.W., Teller, E.: Equations of state calculations by fast computing machines. J. Chem. Phys. **21**(6), 1087–1092 (1953)

91. Mezura-Montes, E., Velázquez-Reyes, J., Coello-Coello, C.: A comparative study of differential evolution variants for global optimization. In: GECCO 06, Seattle, Washington (2006)

92. Morozov, V.A.: Regularization Methods for Ill- Posed Problems. CRC Press, Boca Raton, FL (1993)

93. Mosegaard, K., Tarantola, A.: Chapter Probabilistic approach to inverse problems. In: International Handbook of Earthquake and Engineering Seismology, vol. A, pp. 237–265. Academic, London (2001)

94. Moura Neto, F.D., Silva Neto, A.J.: An Introduction to Inverse Problems with Applications. Springer, New York (2012)

95. Narasimhana, S., Vachhani, P., Rengaswamya, R.: New nonlinear residual feedback observer for fault diagnosis in nonlinear systems. Automatica **44**, 2222–2229 (2008)

96. Odgaard, P.F., Matajib, B.: Observer-based fault detection and moisture estimating in coal mills. Control Eng. Pract. **16**, 909–921 (2008)

97. Ogata, K.: Modern Control Engineering. Prentice-Hall, Englewood Cliffs, NJ (1998)

98. Parker, R.L.: Geophysical Inverse Theory. Princeton University Press, Princeton, NJ (1994)

99. Patton, R.J., Chen, J.: A review of parity space approaches to fault diagnosis. In: IFAC SAFEPROCESS Symposium (1991)

100. Price, K.V., Storn, R.M., Lampinen, J.A.: Differential Evolution: A Practical Approach to Global Optimization. Springer, Berlin (2005)

101. Puris, A., Bello, R., Herrera, F.: Analysis of the efficacy of a two-step methodology for ant colony optimization: case of study with TSP and QAP. Expert Syst. Appl. **37**(7), 5443–5453 (2010)

102. Qin, A., Huang, V.L., Suganthan, P.N.: Differential evolution algorithm with strategy adaptation for global numerical optimization. IEEE Trans. Evol. Comput. **13**(2), 398–417 (2009)

103. Ramos, F.M., Velho, H.F.C.: Reconstruction of geoelectric conductivity distributions using a first-order minimum entropy technique. In: Proceeding of the 2nd International Conference on Inverse Problems in Engineering: Theory and Practice, vol. 2, pp. 195–206. Le Croisic, France (1996)

104. Sacco, W.F., Oliveira, C.R.E.: A new stochastic optimization algorithm based on particle collisions. In: 2005 ANS Annual Meeting, Transactions of the American Nuclear Society (2005)

105. Sacco. W.F. Oliveira, C.R.E., Pereira, C.M.N.A.: Two stochastic optimization algorithms applied to nuclear reactor core design. Prog. Nucl. Energy **48**(6), 525–539 (2006)

106. Samanta, B., Nataraj, C.: Use of particle swarm optimization for machinery fault detection. Eng. Appl. Artif. Intell. **22**(2), 308–316 (2009)

107. Seckiner, S., Eroglu, Y., Emrullah, M., Dereli, T.: Ant colony optimization for continuous functions by using novel pheromone updating. Appl. Math. Comput. **219**, 4163 (2013)

108. Shah-Hosseini, H.: The intelligent water drops algorithm: a nature-inspired swarm-based optimization algorithm. Int. J. Bio-Inspired Comput. **1**(1/2), 71–79 (2009)

109. Sharkey, A.J.C., Sharkey, N.E., Gopinath, O.C.: Diverse neural net solutions to a fault diagnosis problem. Tech. rep., Department of Computer Science, University of Sheffield (1999)

110. Shelokar, P.S., Siarry, P., Jayaraman, V.K., Kulkarni, B.D.: Particle swarm and ant colony algorithms hybridized for improved continuous optimization. Appl. Math. Comput. **188**(1), 129–142 (2007)

111. Silva Neto, A.J., Becceneri, J.C., Campos Velho, H.F. (eds.): Computational Intelligence Applied to Inverse Problems in Radiative Transfer. EdUERJ, Rio de Janeiro, Brazil (2016)

112. Silva Neto, A.J., Lugon Jr., J., Soliro, F.J.C.P., Biondi Neto, L., Santana, C.C., Lobato, F.S., Steffen Jr., V., Campos Velho, H.F., Souza, A.F., Camara L, D.T., Assis, E.G., Silva, F.M., Oliveira, G.P., Camps Echevarría, L., Llanes-Santiago, O.: Direct and Inverse Problems with Applications in Engineering – Research Collection. InTech, Rijeka, Croatia (2016)

113. Silva Neto, A.J., Llanes-Santiago, O., Silva, G.N. (eds.): Mathematical Modelling and Computational Intelligence in Engineering Applications. Springer, Cham (2016)

114. Simani, S., Patton, R.J.: Fault diagnosis of an industrial gas turbine prototype using a system identification approach. Control Eng. Pract. **16**(7), 769–786 (2008)

115. Simani, S., Fantuzzi, C., Patton, R.J.: Model-Based Fault Diagnosis in Dynamic Systems Using Identification Techniques. Springer, New York (2002)

116. Smith, C.R., Grandy, W.T.: Maximum-Entropy and Bayesian Methods in Inverse Problems, Fundamental Theories of Physics. Springer, Dordrecht (1985)

117. Snieder, R., Trampert, J.: Inverse Problems in Geophysics. Samizdat Press (1999)

118. Socha, K.: Ant colony optimization for continuous and mixed-variable domains. Ph.D. thesis, Université Libre de Bruxelles (2008)

119. Socha, K., Dorigo, M.: Ant colony optimization for continuous domains. Eur. J. Oper. Res. **185**(3), 1155–1173 (2008)

120. Sorsa, T., Koivo, H., Koivisto, H.: Neural networks in process fault diagnosis. IEEE Trans. Syst. Man Cybern. **21**(4), 815–825 (1991)

121. Sousa, F.L.: Generalized extremal optimization: a new stochastic algorithm for optimal design. Ph.D. thesis, Graduate Program in Applied Computation, Instituto Nacional de Pesquisas Espaciais, INPE-9564-TDI/836 (2003). (In Portuguese)

122. Souto, R.P., Stephany, S., Becceneri, J.C., Campos Velho, H.F., Silva Neto, A.J.: On the use of the ant colony system for radiative properties estimation. In: 5th International Conference on Inverse Problems in Engineering – Theory and Practice (V ICIPE), vol. 3, pp. 1–10. Leeds University Press, Leeds, Inglaterra (2005)

123. Storn, R., Price, K.: Differential evolution: a simple and efficient adaptive scheme for global optimization over continuous spaces. Tech. Rep. 12, International Computer Science Institute (1995)

124. Storn, R., Price, K.: Differential evolution: a simple and efficient adaptive heuristic for global optimization over continuous spaces. J. Glob. Optim. **11**(4), 341–359 (1997)

125. Suganthan, P., Hansen, N., Liang, J., Deb, K., Chen, Y.P., Auger, A., Tiwari, S.: Problem definitions and evaluation criteria for the CEC 2005 special session on real parameter optimization. Tech. rep., Nanyang Technological University (2005)

126. Tarantola, A.: Inverse Problem Theory. Elsevier, New York (1987)

127. Tarantola, A.: Inverse Problem Theory and Model Parameter Estimation. SIAM, Philadelphia, PA (2004)

128. Tarantola, A., Valette, B.: Inverse problems=quest for information. J. Geophys. **50**(3), 159–170 (1982)

129. Tvrdik, J.: Adaptation in differential evolution: a numerical comparison. Appl. Soft Comput. **9**(3), 1149–1155 (2009)

130. Venkatasubramanian, V., Rengaswamy, R., Yin, K., Kavuri, S.N.: A review of process fault detection and diagnosis-part I: quantitative model-based methods. Comput. Chem. Eng. **27**(3), 293–311 (2003)

131. Venkatasubramanian, V., Rengaswamy, R., Yin, K., Kavuri, S.N.: A review of process fault detection and diagnosis-part II: qualitative model-based methods and search strategies. Comput. Chem. Eng. **27**(3), 313–326 (2003)

132. Venkatasubramanian, V., Rengaswamy, R., Yin, K., Kavuri, S.N.: A review of process fault detection and diagnosis-part III: process history based methods. Comput. Chem. Eng. **27**(3), 327–346 (2003)

133. Wang, L., Niu, Q., Fei, M.: A novel quantum ant colony optimization algorithm and its application to fault diagnosis. Trans. Inst. Meas. Control **30**(3/4), 313–329 (2008)

134. Willsky, A.S.: A survey of design methods for failure detection systems. Automatica **12**(6), 601–611 (1976)

135. Witczak, M.: Advances in model based fault diagnosis with evolutionary algorithms and neural networks. Int. J. Appl. Math. Comput. Sci. **16**(1), 85–99 (2006)

136. Witczak, M.: Modelling and Estimation Strategies for Fault Diagnosis of Non-Linear Systems From Analytical to Soft Computing Approaches. Springer, New York (2007)

137. Wolpert, D., Macready, W.: No free lunch theorems for optimization. IEEE Trans. Evol. Comput. **1**, 67–82 (1997)

138. Yang, X.S.: Firefly algorithm, stochastic test functions and design optimisation. Int. J. Bio-Inspired Comput. **2**(2), 78–84 (2010)

139. Yang, X.: Nature – Inspired Optimization Algorithms. Elsevier, Amsterdam (2014)

140. Yang, X.S., Deb, S.: Cuckoo search via Levy flights. In: World Congress on Nature and Biologically Inspired Computing (NaBIC 2009), pp. 210–214. IEEE Publications, Piscataway, NJ (2009)

141. Yang, X.S., Deb, S.: Engineering Optimisation by Cuckoo Search. Int. J. Math. Model. Numer. Optim. **1**(4), 330–343 (2010)

142. Yang, S.H., Chen, B.H., Wang, X.Z.: Neural network based fault diagnosis using unmeasurable inputs. Eng. Appl. Artif. Intell. 1 **13**(31), 345–356 (2000)

143. Yang, E., Xiang, H., Guand, D., Zhang, Z.: A comparative study of genetic algorithm parameters for the inverse problem-based fault diagnosis of liquid rocket propulsion systems. Int. J. Autom. Comput. **4**(3), 255–261 (2007)

144. Yang, Z., Tang, K., Yao, X.: Self-adaptive differential evolution with neighborhood search. In: IEEE Congress on Evolutionary Computation (CEC2008), Hong Kong, pp. 1110–1116 (2008)

145. Yew-Soon, O., Meng-Hiot, L., Ning, Z., Kok-Wai, W.: Classification of adaptive memetic algorithms: a comparative study. Syst. Man Cybern. Part B: Cybern. **36**(1), 141–152 (2006)

146. Zaharie, D.: Influence of crossover on the behavior of differential evolution algorithms. Appl. Soft Comput. **9**(3), 1126–1138 (2009)

147. Zhan, Z.H., Zhang, J., Li, Y., Shi, Y.H.: Orthogonal learning particle swarm optimization. IEEE Trans. Evol. Comput. **15**(6), 832–847 (2011)

148. Zhang, J.: Adaptive differential evolution with optional external archive. IEEE Trans. Evol. Comput. **13**(5), 945–958 (2009)

Printed in the United States
By Bookmasters